T0071797

Also by Steven Stoll

The Fruits of Natural Advantage:
Making the Industrial Countryside in California

Larding the Lean Earth

Larding the Lean Earth

SOIL AND SOCIETY IN NINETEENTH-CENTURY AMERICA

Steven Stoll

HILL AND WANG
A DIVISION OF FARRAR, STRAUS AND GIROUX
NEW YORK

Hill and Wang
A division of Farrar, Straus and Giroux
19 Union Square West, New York 10003

Printed in the United States of America
Published in 2002 by Hill and Wang
First paperback edition, 2003

The Library of Congress has cataloged the hardcover edition as follows:
Stoll, Steven.
Larding the lean Earth : soil and society in nineteenth-century America / Steven Stoll.
p. cm.
Includes bibliographical references (p.).
ISBN 0-8090-6431-6 (hc)
1. Soil conservation—United States—History—19th century. 2. Soil fertility—United States—History—19th century. 3. Agriculture—Environmental aspects—United States—History—19th century. 4. Land settlement—United States—History—19th century. I. Title.

S624.A1 S77 2002
631.4'5'097309034—dc21

2002023279

Paperback ISBN 0-8090-6430-8

Designed by Jonathan D. Lippincott

www.fsgbooks.com

1 3 5 7 9 10 8 6 4 2

For my parents

Our woods have disappeared, and are succeeded, too generally, by exhausted fields and gullied hills. The land of our ancestors which nourished our infancy, and contains the bodies of our fathers, must be improved or abandoned.

—*Memoirs of the Society of Virginia for Promoting Agriculture* (1818)

Nothing in nature is exhausted in its first use. When a thing has served an end to the uttermost, it is wholly new for an ulterior service. In God, every end is converted into a new means.

—Ralph Waldo Emerson, *Nature* (1849)

Contents

Acknowledgments

My greatest institutional debt is to Yale University. I owe special gratitude to History Department Chairs Robin Winks and Jon Butler and to Richard Brodhead, Dean of Yale College, for a Morse Fellowship in the Humanities during the year 1999–2000, when I researched and wrote most of this book. I also received two generous awards from the Whitney Humanities Center at Yale University: an A. Whitney Griswold Faculty Research Grant and a grant from the Frederick W. Hilles Publication Fund to cover the cost of acquiring and producing illustrations. Nancy Godleski, the American specialist on the reference team at Sterling Memorial Library, worked on my requests for books and microfilm material. I depended on the Records of Antebellum Southern Plantations on microfilm, and Kevin Pacelli made using the Microform Reading Room pleasant and easy. I also wish to thank the staffs of the Seeley G. Mudd Library and the Government Documents Center.

I am grateful for a Kate B. and Hall J. Peterson Fellowship for research in residence at the American Antiquarian Society during the summer of 1998. The staff of AAS is as remarkable as its collections. I especially thank Joanne D. Chaison, Babette Gehnrich, and Marie E. Lamoureux for their knowledge and kindness. This book also depends on the collections of the University of Pennsylvania, the Library Company of Philadelphia, the Pennsylvania Historical Society, and the Winterthur Library in Delaware, where a librarian whose name I never learned agreed to copy and mail to me the complete journal of James Pemberton Morris. Thank you!

Colleagues at Yale and other institutions read the manuscript, including John Mack Faragher, Doron Ben-Atar, Jack Temple Kirby, and Colin Duncan. Two anonymous readers provided me with some of the most detailed comments I received. George Miles helped me with illustrations. Robert Forbes shared his immense knowledge of the early Republic.

I delivered portions of the book as papers to the American Society for Environmental History and the Society for the History of the Early American Republic. I would like to thank James Scott for inviting me to lead a discussion at the Agrarian Studies Seminar at Yale University and Scott R. Nelson for his comments. Deborah Fitzgerald invited me to the Massachusetts Institute of Technology for a productive seminar and an even better dinner. Michael Keilty, a member of the Sustainable Agriculture Program at the University of Connecticut, took the time to show me around his lovely farm. David and Elsie Kline opened their home to me and let me wander around. Richard Meinert patiently served as my soil and manure adviser, though any errors of fact are my own. Curt Ellis introduced me to David Kline's books.

Hill and Wang is a great press because it has Lauren Osborne as its executive editor. This book has benefited from Lauren's exacting standards and fine judgment. Both she and Catherine Newman moved the manuscript through its stations with skill and pleasantness.

Before the book there was its title. Randy Anderson quoted to me Prince Henry's description of Falstaff in *King Henry IV, Part 1* (act 2, scene 2) and suggested that it might be of use: "Away, good Ned. Falstaff sweats to death, / And lards the lean earth as he walks along."

My wife Sara and our children, Batsheva, Katya, and Elijah, made working at home even more fun than the library.

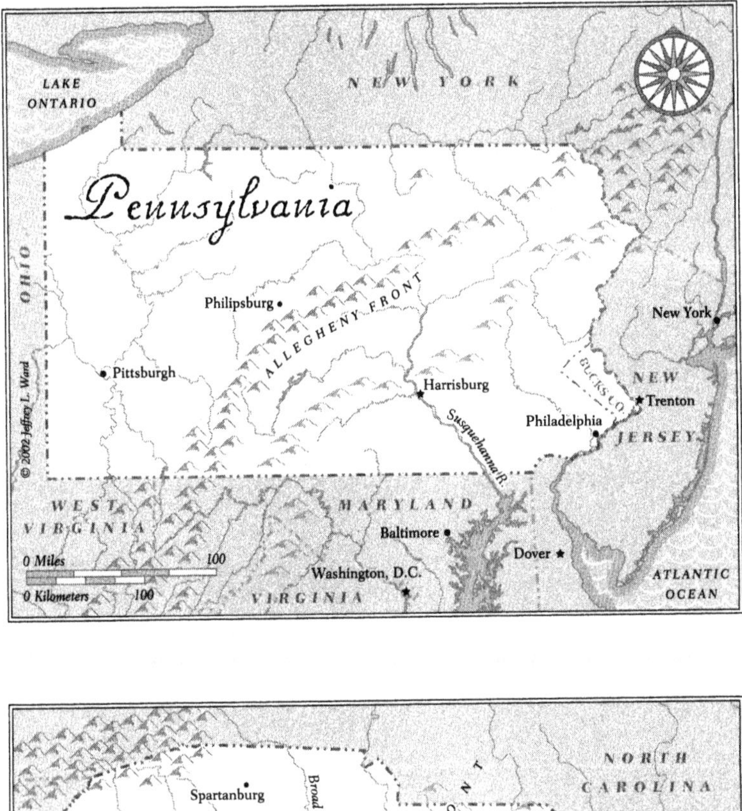
LAKE ONTARIO
NEW YORK
Pennsylvania
OHIO
Philipsburg
ALLEGHENY FRONT
New York
Pittsburgh
Harrisburg
BUCKS CO.
NEW JERSEY
Trenton
Susquehanna R.
Philadelphia
© 2002 Jeffrey L. Ward
WEST VIRGINIA
MARYLAND
Baltimore
Dover
0 Miles 100
0 Kilometers 100
Washington, D.C.
VIRGINIA
ATLANTIC OCEAN

NORTH CAROLINA
Spartanburg
Broad R.
SOUTHERN PIEDMONT
Pee Dee R.
Darlington
Saluda R.
Columbia
South Carolina
GEORGIA
Savannah R.
Macon
Charleston
ATLANTIC OCEAN
0 Miles 100
0 Kilometers 100
Savannah
© 2002 Jeffrey L. Ward

Larding the Lean Earth

Prologue: Litchfield

I have no rubber boots. As we stand in the commuter parking lot at the intersection of Route 8 and Route 118 in Litchfield County, Richard Meinert regards my shoes. Meinert is an expert in animal manure at the University of Connecticut's Cooperative Extension Service, employed to advise farmers on matters of soil fertility and waste disposal, but his only concern at this moment is my getting soaked to the socks in filth. We take off for the center of town in his truck, talking phosphorus and sandy loams. Phosphorus is one of the nutrient elements essential for plant growth, and when farmers run out of it they buy it from quarries in Florida. After corn rich in Florida phosphorus rises on the Illinois prairie, it is sold to New England dairy farmers, who feed it to their Holsteins. The cows void what they can't use. Now at least three times displaced, the phosphorus is discharged into the watershed of the Connecticut River Valley, where it leaches through grainy tilth to foul streams and groundwater. There are all sorts of overlapping cycles linking agriculture to the rest of nature, not all of them beneficial. Soil fertility now depends on far-flung networks and can be implicated in the debasement of ecological systems, but this was not always so. The renewal of nutrient elements once took place within individual farms, conducted through an elegant orchestration of soils, plants, and animals.

We are driving through northwestern Connecticut, a kind of basin and range in which timothy meadow is punctuated by forested basaltic rock ridges. This is a working countryside, with the striking presence of a place more distant than a mere forty miles from Hartford. Hundreds of millions of urban and suburban consumers on the other end of the farm economy have no idea that without people like Richard Meinert the technological food chain that keeps them alive would disintegrate, and he is always reminded of it. "I've listened to parents talk to their kids at fairs . . . and tell them, 'We don't eat those cows, we eat the cows that come from the grocery store.' " Who better than a manure expert to measure the gulf of incomprehension between shopping cart and pasture? Kids today, he said, "they think that chocolate milk comes from brown cows." I've been feeling a little out of touch myself. Months of reading and writing about the ecological ideas implicit in old American husbandry have left me without a tactile sense of the world I want to depict. I have been consulting with Richard Meinert on and off for a number of months about how farmers and planters just after the American Revolution might have gone about the more earthy tasks they almost never wrote about. To make sense of what I only knew from documents, I asked for a ground check, and the good-natured expert agreed to take me along on two farm visits.

We turn off the two-lane highway and onto the gravel drive of Michael Keilty's farm. The proprietor is there to greet us and immediately shows us a shed with dried medicinal plants and herbs for sale. We are all testing for rapport, and although he doesn't say so, Keilty clearly wants to know what I have come so far to see. I thank him for the lesson in botany and ask to see the barn. Keilty has a salt-and-pepper beard, and he's comfortable and easy in wire rims that make him look professorial. He was born in this county, not ten miles from where his grandparents and great-grandparents once farmed. He left home to take degrees in agriculture and wildlife management and to teach for a few years. Then he came back to plow it all in. Everything old and valuable has a provenance, and the region including these fifty-five acres came under European control when King George III granted it to the Episcopal Church in 1747. It came into Keilty's hands overgrown and weedy with invasive species. Clearing it, planting it, making it pay have been his work for twenty-five years.

Looking straight through the barn, I see meadow in the foreground and into the distance. Between the daylight all around and the dim of old milking stalls stands an enclosure with about fifty sheep. They rump and bump and step over each other as we approach to take a closer look. The pen consists of straw and dung in a layer about eight inches deep. The animals wallow here for as long as six months before Keilty cleans it out. Microorganisms take up the volatile urea, turning it into an available form of nitrogen so that the dung becomes a kind of slow-release capsule when it reaches the soil, and they give off enough heat to warm the barn in winter. The compost is then food for vegetable beds, forage crops, and the meadow, which Keilty seeds every year in timothy and other grasses. Seven cows camp in the open air next to a hay feeder, which Keilty moves every few weeks to spread the dung. Thirteen acres of meadow will not be grazed until the sod thickens and the timothy forms heads of soft seed. The rye in long beds near the house serves a different purpose. It holds nutrients safe from wind and water until it is turned under, when its nutrients will return to the ground in preparation for vegetables. The beds never stay in the same place for long but revert to other uses every few years. Keilty's methods are not complicated. They are not new. They are not easy to practice on farms larger than about two hundred cultivated acres, and for that reason they are not very popular today among American farmers. What makes them important is that they complete a cycle.

Food production that pays its debt to soil is not simply a matter of injecting inputs at one end and harvesting at the other. It requires a farmer to usher nutrients through the seasons and weave them through plants, digestive systems, and soils. It requires the constant addition of organic matter. The way in which Keilty shifts the location of his beds and meadows, treating them as temporary uses within a larger system, one that never needs fresh land, would have been instantly recognizable to a small number of American farmers in the 1820s.

Climbing over a fence rail on our way back to the house, I trip over a pack of dogs that have urged us on to the meadow and back. One of my shoes (a respectable utility oxford with a Vibram sole) goes flying off into the turf, and I stumble as the dogs bound for the house and leave me hobbled. We all laugh at me as I return my damp foot to my pitiable, dung-covered shoe. I can't help stepping on lots of black sheep clods all the rest of the way. The essence of what I call restorative husbandry is giving me a little trouble, and I have lost some of the desire to embrace my subject, but my visit has somehow inspired Keilty. "My grandmother referred to manure as like having diamonds. The more manure you had, the richer crops you could grow; the more crops you could grow, the more livestock you could have." Think of that! An *upward* spiral in affluence and stability from a long-labored soil, land restored and not degraded by unceasing food production, simple practices promising sustainability—these we do not ordinarily associate with the environmental history of North America.

That story is all about greed and waste. Europeans arrived tens of thousands of years after the ancestors of the American Indians and went about dividing land, cutting trees, damming rivers, shooting buffalo, ripping up mountains for minerals, killing off whales, and skimming fertility over and over again in every new place. The free-for-all continued until a few people warned that game animals and timber supplies would not recover without human intervention. The United States put these fears into public policy by nationalizing millions of acres, mostly in the West, in order to regulate the private use of resources, and that became "conservation." Yet this is only one thread of a richer story, and it has nothing to say about Michael Keilty's grandparents. Animal manure anchored a nutrient cycle with the potential to create a steady state between taking away and putting back. Not only that: when farmers in the past husbanded soils in addition to plants and animals, they installed a system that never needed to be fed with fresh land. In short, these practices conserved not only fertility at the farmstead but also a greater landscape of forests and lowlands; and by allowing families to stay rather than migrate overland, they offered stability to rural communities. Historians have recently begun to tell a more complicated story of conservation, with attention, for example, to the resistance of people living in places where the government imposed boundaries and regulations that had never existed before, but farmers are often not included in these stories.

The "improvement" of agriculture represented an important event not only in the history of North America but in the history of civilization. Following a three- or four-thousand-year period of domestication, which took place in locations throughout the world beginning ten thousand years ago, little changed in the ways people kept their animals and tended their crops until the sixteenth century. A number of medieval inventions, like the horse collar and the three-field system of rotation, certainly qualify as notable, but the methods that finally reached their highest expression in England in the 1750s—which I explain in great detail in their place—amounted to an infusion of intellectual energy unlike any other since domestication itself. This book makes shooting glances back—way back—to the first Neolithic practices, because I see the changes that emerged in Britain and the United States as consequential in a longer history. Placed next to the long view, the short period between the War of 1812 and the California Gold Rush may not seem like much to get excited about. It was just a moment, but one worth our understanding, for in the moment before the Far West became part of the United States and before railroads and full-scale urban industrialism took off, a group of farmers imagined a process of production that also brought about the reproduction of their land. In 1820s America it still seemed possible to contain the bad tendencies of consumption within a delicate system of return, powered by the sun and managed by cultivators who saw soil as the totality of matter passing through their hands.

Some readers might regard the 1820s as distant enough to be ancient, but at least one concern from that time has recently become central to the concerns of our own. The most perplexing environmental problems can no longer be blamed on reckless corporations dumping their refuse into rivers or cutting forests outside the law. Instead, they arise from the sum of individual decisions about consumption and disposal, which cannot be monitored as easily as the pollutants leaving a factory smokestack, resulting in astonishing changes to the Earth. When it comes to the external consequences of how we live day to day, many people in developed countries refuse to draw a line between liberty and responsibility. It may be our liberty to create solid waste for landfills or to load carbon into the atmosphere by driving long distances, but when does it become our responsibility to act differently whenever possible? The story of agriculture early in the nineteenth century is not about environmentalism, a twentieth-century political term if ever there was one, but it is about consumption and its consequences at a time when individuals could not and would not be held accountable for their abuse of land. The farmers I highlight recognized limits where others did not. With the exception of American Indians, who did some of the same but differently, no one in American history preceded them in these ideas.

If I have a single point to make with this book, it is that farming matters. It is the central biological and ecological relationship in any settled society and the most profound way that humans have changed the world over the last ten thousand years. Farming defines a specific landscape, the middle landscape—that place somewhere between wilderness and city where settled societies produce all of their food. Farms are where people engage in aggressive manipulations of plants and animals and also where they learn the limits of what they can take from nature. Farming occupied the vast majority of Americans in the nineteenth century, tied up most of the capital, and created the most essential commodities. The environmental history of North America is unintelligible without agriculture because husbandry embodied the force of settlement, created cultural landscapes, sustained the entire population, and produced commodities for trade and manufacturing. Environmental change did not cease after land had been cleared and farmed for a generation. Deforestation, erosion, the destruction of habitat—farmers caused them all. Historians have tended to overlook the mucky detail of this story for the simple reason that they have not known how to read it. They have slighted the dirt under the fingernails of rural life even though practice and process were fundamental to writing about the countryside during the early nineteenth century.

I have been digging in the mucky detail. I want the reader to see cultivation as a crucial force in the environment and farmers as the most important conservationist thinkers in the United States before the 1850s. I do not write about a single state or region, nor do I dwell on any specific piece of landscape for very long. I emphasize Pennsylvania and South Carolina because those two states form a contrast and allow comparison between farmstead and plantation. My working assumption is that the ways in which people derive their livelihood open a view to larger trends in politics, economy, society. What involves me most is a short span of years when the cycling of nutrients through soils, animals, and crops became synonymous with progress. Contained in the smells and chemical reactions of the barnyard is an entire history of the American environment told from ground level. Improvement blended ecology with ideology, practice with politics, nature with the future of the Republic.

Meinert and I left Michael Keilty in his broad-beamed house after exchanging handshakes and e-mail addresses. Later we visited a modern dairy operation with cement floors and an enclosed milking parlor for hundreds of cows, where I saw a pool half an acre large filled with liquid effluent. The smell hit me like a wave of heat. Food production is fraught with many more messy questions than millions of consumers care to consider. Strangely, what comes out of the other end of a cow opens a dynamic view to the environment.

This book is written in three long chapters, each divided into sections of varying lengths. Chapter One considers the idea of improvement and its context in the political economy of the early Republic. Chapter Two visits the problem of restoration and reform in two states from the 1820s to the 1840s. My intense focus on only a few farmers and planters allows me to write larger regional stories around them, based on an array of printed sources. The argument concludes in Chapter Three with the legacies of rural reform. Landscape gardening, irrigated settlement in the Far West, and the rush of mechanical and chemical innovation that evolved into industrial agriculture link back to the farmers of the 1820s. Most significant of all, improvement fed a certain current that became conservation, signified by the life and work of the scholar and statesman George Perkins Marsh. Whereas most books about conservation begin with Marsh and his seminal *Man and Nature; or, Physical Geography as Modified by Human Action* (1864), this one concludes with him as the product of a prior movement. Yet these chapters do not offer a perfect explanation for the demise of improvement. The energy that fed it was conserved, but it can never be recaptured in its earlier form.

One

Let us boldly face the fact. Our country is nearly ruined.

—John Taylor (1819)

The times are changed; the face of the country is changed; the quality of the soil has changed; and if we will live as well, and become as rich and respectable as our fathers, we must cultivate their virtues; but abandon their system of farming. —*The Farmer's Manual* (1819)

The changes, which these causes have wrought in the physical geography of Vermont, within a single generation, are too striking to have escaped the attention of any observing person, and every middle-aged man who revisits his birth-place after a few years absence, looks upon another landscape than that which formed the theatre of his youthful toils and pleasures. —George Perkins Marsh (1847)

Forming the Furrow Slice

Agriculture is an act intimate with the rest of nature. Like the apparent separateness of species in an ecosystem, the boundaries of any farm belie a deeper connectedness. Agroecology is akin to the study of similar triangles, in which the smaller figure shares the proportions of the larger. Take water's path through the biosphere. Small-scale transpiration from roots to leaves involves the entire hydrology of a region, the shape and altitude of its watersheds, the granulation and retention of its soils, and, finally, its climate. Porous borders between human-managed and wild nature appear at every turn. Farms cut out of forests retain all sorts of buried seeds waiting for fire or drought before they can germinate. Deer like to graze the young shoots at edges and corners between clearings and canopies. With all of these relationships unfolding at the same time, where should we stand for a view of cultivation as a subset of the natural relationships in any environment? The most vital place lies between what farmers could control and what they could not, between plants and animals on the one hand and geology and climate on the other. Soil.

As the first and most immovable resource of agriculture, soil was the country. Families carried their corn and hogs long distances, but soil was what they found where they set down and sometimes what they confronted. They might take from it or add to it by treating its fertile nutrients as a rotating fund, but they could change its essential characteristics—emerging from moisture, elevation, and the entire biota—only within tight margins. As a subject for interpretation, soil is rich in material. It formed the living tissue between economy and ecology, the plane that unified food production with events in the greater environment. Early in the nineteenth century, soil became the focal point for a conception of nature as strictly limited.

The only fertile soil is topsoil, a layer of black loam just two feet deep where all cultivation takes place. All the plants known to dry land grow in topsoil, so it creates most of our food and oxygen. It is the flash point for everything on Earth and unifies, in one body, the three great spheres of life—the gases of the atmosphere, the minerals of the lithosphere, and the organisms of the biosphere. Topsoil is ever being renewed. Pause at any road cut to observe a narrative of biochemical reactions and constant weathering. At your feet are deep deposits of parent rock that blend into higher levels made up of granular minerals, including silicate clays, iron and aluminum oxides, gypsum, or calcium. This makes up subsoil. Just above, in a stratum black in color, soft to the touch, and dense with life, is topsoil. Roots braid through it, and worms and insects tunnel around in it, making it spacious and pliable. Microorganisms consume its dead plants and animals, break them down to basic nutrients, and in this way turn tissue back into plant food. A single gram of it might contain billions of bacteria, a single square foot might host a multitude of bugs and fungi. Topsoil holds most of the available water in any ecosystem. Without this reservoir, moisture finds the nearest watercourse; land dries out; climate changes. It is a filter and a container, a mass of integrated micro and macro matter, and a living substance that cannot be understood by reduction. Its final form contains so many members and symbiotic relationships that it constitutes, in the words of the soil scientist Nyle Brady, the "genesis of a natural body distinct from the parent materials from which the body was formed."[1] Soil is the tablecloth under the banquet of civilization: no matter what people build on it, when it moves all the food and finery go crashing. It is the skin of the Earth and is so completely associated with cultivation that it takes a name descriptive of the very act of opening land: the furrow slice.

Farmers gave it names that reflected their work in it, part of a nearly forgotten language of husbandry. There is no better word than *tilth* to illustrate agriculture as a soil practice. In its earliest, Old English form, with a first recorded usage dated to 1023, *tilth* referred to any labor applied to land for subsistence. Since work indicates the thing being worked upon, *tilth* soon described farmland under cultivation and the condition of that land. Worked land was land in tilth, or simply tilth. Even more graphically, *tilth* described the feel of a prepared surface, the depth of it over the top of a boot, the way it fell away from the leading edge—the crumb. Fragrant and deep, with the consistency of butter or stiff dough, tilth defies description as dirt. Why all this attention? Because land includes so much out of human control—the bedrock, the blue sky—and *tilth* names the one part of land most directly a farmer's burden and responsibility.

Tilth implies the plow, maker of the slice. The plow is an implement used to open the ground by turning it over to form long trenches called furrows. It is a frame with a blade or stick that cuts while being pushed or pulled. There seems to be no recorded reference to the plow in English before 1100, but soon thereafter it began to leave its imprint. It has been *sulh* and *ploh* and *ar*, as in *arable*. Its spelling could not be pinned down before the sixteenth century, with some writing it *pleuche* or *plwch* or *pliff*. *Plough* or *plow* has been in use only since about 1700. It can be combined with a hundred words for various meanings, all related to clay and clod: *plough-feast*, *-harness*, *-mark*, *-rein*, *-servant*, *-team*, *-wheal*, *-woman*. In some regions of England plough-boys collected plough-penny on Plough-Monday, when a tenant was bound to plough for his lord. Like tilth, land took the name of the action upon it, as in "they stepped out of the yard and walked through the fresh plow." The plow is at once the symbol of domestication and the world's most feared ecological wrecking ball. It unearths micro environments, destroys nests and burrows, throws open moisture. Whether they accept the charge or not, plow-people become the caretakers of soils. They become the roots and the sod and sometimes even the rain. They become the holders of the fertile layer not only because they depend on it but because they can destroy it. Here is a contract with nature that has not always been honored: rip the earth but be there to stop the bleeding.

If agriculture is a controlled disturbance, then churned-up ground is its worst upheaval, carrying the constant threat of fertility lost to crops and the weather. When they took the trees and broke the ground on a steep slope, when they cut furrows in a wet country or in a dry and windy one, farmers brought on a dreaded reaction. Erosion is the ground giving way, and since that ground feeds counties, countries, and continents, erosion suggests a squandered future, a foolish settlement. Tillage makes soil vulnerable to a rate of weathering by wind and water thousands of times faster than the processes that create soil. Erosion takes the ground in two general ways: gullies and sheets. In the first process, water removes soil by creating small channels or rills that become larger gullies as they widen and deepen geometrically until they consume vast areas. With vegetation removed and the surface broken on a slope of greater than five or ten degrees, rain can bring down a hillside within a few years. Even the raindrops themselves do damage. They strike exposed dirt, loosen it, and destroy its delicate granulation. The smaller particles are then easily transported by running water. Badlands result, with pedestal-like formations and no surface for cultivation whatsoever. In the second process, wind and water sweep and skim the surface for years. Fertility falls, crops are flooded or blown away, or their roots are exposed, or they are covered with dust. Barren expanses result, with subsoil laid open.

One Vermont farmer reported the case of the Tinmouth plains, apparently a region of loose soil over hardpan. A rich "muck" ran away down the watercourse within a generation, silt for harbors, leaving him at a loss:

> Twenty-five years ago, I ploughed fields on the Tinmouth plains, which had then a covering of six inches or more of this muck, and they had then been cropped some ten or twelve years; these fields have now not one particle of muck or mold! What has become of it? For the first five to ten years, as I am informed, no fields produced finer crops of wheat than these; the muck has now all disappeared; nothing but the granite sand remains, hardening as we descend, till at one foot or less we get into a perfect hard-pan. . . . Where once the sickle paid its annual visits, it is known no longer.[2]

A soil with little structure, its heavy roots and covering vegetation removed, will slide away, leaving nothing but naked rock.

Exhaustion is another kind of erosion. Although dirt might not fall off the farm, its nutrients can be removed by plants until nothing but poverty grass will grow. As the New York farmer and editor Jesse Buel reminded his readers, "Every crop taken from a field diminishes its fertility, by lessening the quantity of vegetable food in the soil. Unless, therefore, something in the form of manure is returned to the field, an annual deterioration will take place until absolute barrenness ensues." The relationship between profitable yields and the balance of nutrients in soils emerged as one of the most intractable problems in American agriculture before the twentieth century, because, though farmers understood that fertility could be taken, few understood the best way to put it back and almost no one knew the chemistry behind any process. Plants use sunlight to photosynthesize sugars from carbon dioxide and water, and they also synthesize amino acids and vitamins from inorganic elements—especially nitrogen, phosphorus, and calcium—taken up from soil.[3] Domesticated animals provided an escape from this one-way alley leading to depletion because their dung was relatively rich in nutrients and, when properly applied to fields, restored most of what crops removed. Erosion and exhaustion were not always easy to differentiate, however, and farmers rarely made any attempt to tell the difference. Lost fertility is itself a kind of erosion, and some forms of erosion are only perceptible, at first, by declining yields. *Exhaustion* often stood for any noticed deterioration in fertility or the volume of tilth. Cotton planters called land *oldfield*, a term for acres worn out and fit for pines, whether its topsoil had gone to market or down the gully.

People are anchored in place only as securely as the ground they till. Lost soil is unrecoverable, and the pace of its formation is so slow that the end product must be considered nonrenewable. One survey of a southern district in the 1930s found earthworks abandoned on land not cultivated since 1887. Under the pine crowns, on high ground, the researchers found fifty years of accumulated topsoil one-sixteenth of an inch deep. At that rate of creation the pines would see their first inch in eight hundred years, their first foot in ninety-six hundred years—the age of agriculture itself. In human time it can be lost forever. No matter the mode or the cause, where topsoil is removed or degraded agriculture is impossible and a settled people cannot persist. Subsoils are the bones of the earth. They have no living organisms and no rotting plant food, and they hold little water. All these are lost with topsoils, and people follow.[4]

When the hunters of the Fertile Crescent ceased moving from place to place in search of food beginning about ten thousand years ago, they commenced a new relationship with the earth. The domestication of einkorn wheat and two-rowed barley did not alter their commanding position as top predators in the food chain, nor did it stop them from hunting and trading over great distances in an extensive relationship with their environment. But one thing did change. When they settled, humans derived a larger portion of their food from the farmstead. Living in one place, tilling there year after year, brought the possibility of scarcity home. Whereas before the great transition they rid a region of game through overhunting and followed another herd or traveled from a place bearing roots to where they could find fish, they now wedded the source of their subsistence. But agriculture is an uncontrolled experiment. For over 99 percent of human history, people hunted and gathered their calories. Place the past three million years of primate evolution equal to a day, and the age of agriculture begins at 4.8 minutes before midnight. Settled society is so new that we are still trying to understand its basic patterns, and it is not at all clear that fixed cultivation is as well adapted or will last as long as the methods of subsistence it replaced.

Historians have often referred to the civilization of ancient Mesopotamia, embracing the floodplain of the Tigris and Euphrates Rivers in Iraq, as the most foreboding example of how the first settled people jerked the ecological rug from under their own feet. A sophisticated network of canals for irrigation captured waves of silt that could be cleared only with extraordinary labor. The ruin of the canals resulted in the near abandonment of Babylonian cities by the last decades of the eighteenth century B.C. One anthropologist described the region as bleak and barren, "yet at one time here lay the core, the heartland, of the oldest urban, literate civilization in the world."[5] An epic of urban formation followed by dusty abandonment looks menacing next to millions of years of hunting and gathering, as though settled societies are inherently unstable.

It should be said that pre-Neolithic hunters also made trouble for themselves. One theory holds that after leaving Africa about forty thousand years ago, early humans caused the extinction of large-bodied mammals (mastodons, saber-toothed tigers, giant sloths) on every island and continent they called home. People live within limits no matter how they live, but by installing subsistence more firmly in one place, agriculture narrowed the niches, narrowed the species, narrowed the limits. Even the earliest farmers understood that their nourishment and comfort depended on what they could transfer from soils to crops and livestock and that if they wanted nourishment and comfort for more than a generation or two they needed to make a transfer back. Soil is a bank account for fertility that farmers draw upon, and the balance is always low.

AN ETHIC OF PERMANENCE

Americans who cultivated the soils of the seaboard spent the balance down. In a common pattern, farmers who had occupied land for only twenty or thirty years reduced the fertile nutrients in their soils until they could no more than subsist. Either that, or they saw yields fall below what they expected from a good settlers' country and decided to seek fresh acres elsewhere.[6] Forests cut and exported as potash, wheat cropped year after year, topsoils washed—arable land in the old states of the Union had presented the scares of fierce extraction by 1820. Northern farmers fled rocky hillside fields they could no longer maintain by doing things the old way. Modes once necessary, said one reformer before an audience in Philadelphia, "had been unwisely pursued by the successors of the hardy conquerors of the forest long after that necessity had ceased to exist, until at length the once fertile fields of the parents yielded but a mere competence for their children, and afforded but a pittance for the generation that followed." In other words, as one historian wrote, "The Vermonters used up Vermont."[7] Rather than stay and make oldfield provide again, southern planters uprooted with slaves and seeds to Alabama and Mississippi. Throughout the nineteenth century, eastern farmers emigrated—pushed and pulled by a lack they caused but would not replenish.

There is nothing very surprising about Americans on the move. Striking out opened the continent, eventually bringing families to Missouri in the 1820s, to Texas in the 1830s, to Oregon and California in the 1840s and 1850s. Overland families settled up the river valleys in advance of the federal survey; they called for the Army to fight and treat with Indians; they were the speculators and the homesteaders. The people who removed in search of land or minerals left behind one life for another, and it is with them that historians have often traveled, peering into their households and social relations, asking questions about their motives and religions, and following them to where emigrants formed new communities in mining camps and cattle towns and isolated farms. Yet for all its guts and power, the great American emigrant story never carried me away. Instead, I've been struck by the varied collection of people for whom striking out for fresh land represented environmental destruction, moral failure, political decline, and economic disaster. They had a way of looking at the common practices of farming and the movement west that was fundamentally different from that of the emigrants.

The fertility of soil became a source of worry for a group of North American planters and farmers in the decades following the War of 1812. They experimented with agricultural practice and urged their findings through the rural press. Most of all, they recognized a link between an enduring agriculture and an enduring society in the long-settled places and picked up an old word to name their efforts: *improvement*. Improvement makes anything better by raising it from a rude to a more refined state or condition, but in American usage it specifically applied to the condition of land. To improve land brought it into "better account," claimed it from wilderness, made it serve the purposes of livelihood. A public that had witnessed the joining of lakes and oceans through the Erie Canal and the deforestation of vast regions to build cities and a nation of farms expressed little doubt, irony, or hesitation about the mastery of human hands over the world.[8] Taking off trees and plowing up grasslands brought benefits, not harm, to farmers, to the expanding United States, and even to the land itself. Yet it speaks for the complexity of the word that *improvement* came to be used somewhat differently by certain cultivators who fretted about the rapidity and the waste of settlement. In this other sense, *improvement* meant the changes that enabled land to be cultivated in the most prosperous possible way over the longest possible time. It meant to invest in certain crops and animals that, when brought together, formed something unique in the nineteenth century—a highly managed natural environment, an ecology at once profoundly disturbed by humans and made more productive by them. The first sense had nothing to say about the natural environment represented by arable acres; rather, it simply celebrated the happy conversion of forest into farmland. The second sense recommended an enduring occupancy through conservation, pointing toward the creation of stable agro-environments.

Reforming American farmers of the early nineteenth century confronted what they believed to be the depopulation of their counties, states, and the Eastern Seaboard. They reacted by assembling the pieces of a new husbandry. Some historians have called it the Agricultural Revolution, and it was no small thing. It amounted to the most monumental advance in landed practice since the invention of agriculture itself. But the meaning of these events may not be what it seems. It may be logical to assume that the Agricultural Revolution grew like an appendage of the Industrial Revolution. In fact, the methods important to this story preceded the steam engine by a century and spun from forces internal to cultivation (though both owe their emergence to a prior capitalist revolution). Until the middle of the nineteenth century, agricultural progress paid no dues to machinery, and American writers scarcely mentioned implements until the 1830s. Without denying the confluence of rising farm production and mechanization later in the nineteenth century, I want to argue that this first leap forward had nothing to do with industrialism. In essence, and in the words of two historians, British farmers and landlords figured out a way to release "the latent powers of the soil on a scale that was new in human history."[9]

Why did a set of thoughtful cultivators living at the daybreak of American expansion worry so much about land and its inherent limits? What brought them to assert an ethic of conservation? That may sound like a word out of time within a half century of the American Revolution. Conservation did not become a full-fledged program of the federal government until late in the nineteenth century, when Congress and the President set aside various lands—especially forests and extraordinary landscapes like the Yellowstone Valley in Wyoming—that would be managed by new agencies. Elite hunters, the graduates of professional forestry schools, and members of organizations like the Sierra Club made up the citizenry that supported conservation. So what links farmers in Pennsylvania and South Carolina during the 1820s to the same movement? Conservation as a set of ideas for the perpetual use of natural resources is much older than the bureaus established by Theodore Roosevelt. Like Roosevelt, improving farmers of the early Republic proposed a measured exploitation of nature as a way of enriching the present generation while promising the same comfort to citizens unborn. Conservation tended to serve materialist, not idealist, ends. The possibility that rural landscape might embody the harmony they perceived in the rest of nature moved farmers, though most drew harder, straighter lines from soils to their fortunes and politics. Whereas progressives of the twentieth century feared that individual opportunity and industrial primacy would fail along with dwindling forest and mineral reserves, improvers of the early nineteenth century feared that population and their state-centered ambitions would fail along with the rural environment.

Even so, improvement differed in fundamental ways from conservation—the former being a matter more of individual restraint than of centralized regulation. Roosevelt appointed government the owner-manager of the public lands with the authority to control devastating consumption. But improvement in the era of Madison packed no federal law and carried no force but persuasion. It urged private discipline in cultivation and criticized the tendency among farmers to plow up too much land at once, spread thin their seed and labor, cut woodlots too quickly. It asked farmers to build fertility as a way to moral progress, a gracious landscape, and worthy wealth. The interpretation of Jesse Buel reads like Enlightenment ethics: "Individuals, it is true, are but units—yet the aggregation of units makes millions, and the aggregation of individuals constitutes nations. We should all act as though individual example had an imposing influence upon the whole." Sounding like Adam Smith, he advised every farmer to adopt the restorative system to promote his own interests, "and by his example, benefit society."[10] Improvement represented conservation at the level of comportment—not the centralized regulation of resources but the individual regulation of consumption.

Here was a contrary vision of progress. Improvers dreamed of a gradual settlement flourishing in a limited space and often took their inspiration from descriptions of fleshy sheep ruminating in the rising dew of the English countryside. At times they could be envious of shining white cliffs marking land's end. Or, as one writer meditated:

> If a wall, like that of China had been built around "the old thirteen" at the time when they resolved to set up for themselves how different would be their aspect, and how much more highly cultivated, populous, strong and comfortable at this time—But our policy has been, by the prodigal management of our public domain, to set in motion a constant current of emigration, which has not only carried off from the sea-board, all accessions of labor and capital from Europe, but which has drained the old states of their most active and vigorous population.[11]

Take stock of these words. They go right to the core and link agronomy to the politics of a certain agrarian class active in the decades before about 1850. What kind of grumbler would imagine a boundary to continental expansion? What kind of complainer, living in a time when steamboats and railroads had begun to erode the once insurmountable nature of space and time, would imagine away everything west of the Appalachians?

The complainers were editors and subscribers to the rural press. Any list of important journals must include the Albany *Cultivator*, the *Farmers' Register*, the *American Farmer*, and the *Southern Agriculturist*.

Niles' Register and *De Bow's Review*, journals with national profiles and circulations, constantly acknowledged agriculture as the foundation of the American economy. These publications never commanded the attention of a majority of farmers—far from it. Jesse Buel, editor of the *Cultivator*, estimated that the thirty journals published in 1838 reached about 100,000 people in a total farm population of half a million families. Little more is known about the individuals who subscribed to these publications than that they tended to be men, practical farmers, and more affluent than average. Henry Balcom of Oxford, New York, kept a subscription list with the names of 132 local readers of the *Cultivator* between 1839 and 1865. A storekeeper, a bookseller, and a builder took the paper. Political leaders read it; so did some of the richest people in town. Most subscribers held a little property; one owned nothing at all.[12] The population of conscientious farmers also included the authors of hundreds of manuals, addresses, and columns, as well as the thousands of members of agricultural societies from Maine to Mississippi. The subject embraces the famous as well as the anonymous. There were the major southern thinkers, like John Taylor (Virginia planter and author of *Arator*) and Edmund Ruffin (also of Virginia, editor of the *Farmers' Register*), both of whom propagated methods that they hoped would bolster the influence of the southern states and increase the profitability of slavery.[13] Northern improvers knew the highbrow Philadelphia Society for Promoting Agriculture, whose list of members included the mercantile elite in addition to gentlemen farmers. Reform at the furrow eventually jumped its bounds and left an imprint on the larger culture in the works of people who owed something to the farmers who came before them. By the 1840s a few, like the landscape designer Andrew Jackson Downing and the Vermont scholar and statesman George Perkins Marsh, had begun to extend the idea of a permanent, highly managed rural ecology to the whole human landscape.

In sum, the people whose words and works will be central to this book never spoke with one voice and therefore defy definition as a national movement, but they tended to agree on this: heedless expansion represented a threat to the economy and society of the old states.

Farmers and planters of this ilk floated a syllogism made up of two premises leading to a conclusion: if *a* does *b* and if *b* does *c*, then *a* does *c*.

1. PREMISE: (*a*) Improved husbandry (*b*) cycles nutrient elements.
2. PREMISE: (*b*) Any practice that cycles nutrient elements (*c*) ensures the productivity of land into the indefinite future, thus encouraging rural people to remain in the old states.
3. THEREFORE: (*a*) Improved husbandry (*c*) holds farmers from westward emigration.

The farmers and planters who referred to themselves as improvers saw the folly of land-financed expansion in the nutrient bankruptcy of common husbandry. They formulated these premises while thinking through the material foundation of their social interests and political ambitions. Their peculiar view of a nation on the move brought them to look harder than others at the uncertain wedlock between peoples and soils.

Improvement is only meaningful in a rich context. As a history of mere scythes and plows, isolated from ideas and ecological change, agriculture has little significance. That farmers in the past shifted their fields between arable and pasture and used cattle to create manure is an unimportant fact. To ask when and why this process gained currency—how, like a pebble in a pond, it altered the world beyond where farmers labored—what larger purpose it served, problems it solved, and events it offset for the people who employed it, is to ask historical questions. Soil restoration and crop rotation existed in their mature forms long before the United States existed, but they soared in importance during the 1820s. The reasons had to do with that period of stunning maturation that began with the signing of the Constitution and ended with the Compromise of 1850, the early American Republic.

None of the practices discussed here first emerged during the nineteenth century. Major farmers like John Beale Bordley of Maryland and Thomas Jefferson of Virginia acclimatized European crops and rotations to North America immediately after the Revolution, but examples of deliberate experiment on this continent go back to the 1740s, when the Connecticut farmer Jared Eliot published his *Essays upon Field Husbandry*. The same methods can be found in the papers of Enlightenment-minded planters.[14] Their instruction came from Great Britain, where Jethro Tull's *Horse-hoeing husbandry* (1733) marked the zenith of an era of landed experimentation, not its beginning. By asserting that nature, not trade, underwrote the true wealth of nations, French economists of the eighteenth century, known as Physiocrats, focused attention on the productivity of land and must also be considered forerunners. Stepping back a ways, the Roman agrarian known as Cato advised, "Make it an aim to have a big manure pile. Preserve the manure carefully." The idea that animal dung replenishes organic matter in soils had been around for thousands of years by the time English landlords came along, most of whom took too much credit for promoting what others had worked out long before. Jesse Buel humbled the project in 1838: "What we denominate the new system, has long been in operation in the valley of the Po, in Italy,—indeed it seems to have been practiced there by the Romans, in the meridian of their greatness—and in Flanders,—and for the last half century in Great Britain."

Yet the political setting that lent new urgency to old ideas had never existed before. Consider just a sample of the accelerating pace of events Americans had to absorb beginning in the 1790s: a proliferation of capitalist institutions like joint-stock corporations and lending banks, the rise of factories and wage work, the increasingly common experience of market transactions and entrepreneurship. Then came horrifying Indian wars in the Northwest, the Whiskey Rebellion (1794), the Alien and Sedition Acts (1798), and the Louisiana Purchase (1803), which doubled the size of the United States. An old order in which agriculture formed the uncontested bedrock of political economy seemed threatened by events arising between the War of 1812 and the end of the war with Mexico. The first war invited the quick and shaky rise of a manufacturing sector, soon to be crushed by a decade-long depression following the Panic of 1819. The depression led into an uproarious tariff controversy in 1827 and 1828, whose outcome set off the Nullification Crisis between South Carolina planters and the federal government under Andrew Jackson. The second war resulted in a continental nation. Gold was discovered near Sacramento, California, in 1848, only months after the American takeover. The ensuing rush for wealth made eastern farmers dispensing advice about cattle and legumes sound like cranky elders. The reformation of agriculture had its moment, and these events—especially the depression, the tariff, and the Gold Rush—defined that moment. Still, the context of our subject is far from clear.

Improvement found voice and gathered adherents during the 1820s because it offered, among other things, an intensified production to match the declining value of commodities and real estate after the Panic of 1819. It offered methods for pulling greater product and profit from fixed limits. From here it might make sense to cast improvement as a creature of the market economy and argue that its many disciplines reflected the values that would soon characterize industrialism, but that is inadequate. In fact, improvement appealed for reasons other than rewards from the market. As the historian James Henretta wrote, describing an earlier period: "A convincing interpretation of northern agriculture must begin . . . not with an ascribed consciousness but rather with an understanding of the dimensions of economic existence." In the words of two economic historians, "A high rate of return was one goal among many." Merchants became rich by building capital reserves, yet even those farmers engaged in commercial production reinvested everything in their property and remained adherents of a husbandry "not carried on as a means of making money, but rather as a mode of existence."[15] It overstates things even to say that farmers *made money*. They earned little more than 6 percent, often closer to 3, during the 1830s and 1840s. Rates of return from manufacturing often reached double those numbers, but farmers continued to plow labor and meager capital back into their homesteads. They had their reasons. Most would never own the minimum financing to enter a manufacturing business; they enjoyed a lower financial risk compared with those engaged in other occupations; and, as it turns out, they best employed labor and meager capital when they improved soil and buildings. These fixed assets increased the value of the homestead and brought the surest profit if and when a family sold. Certain farmers and many planters accumulated wealth, but the majority pursued their happiness, in the words of one Pennsylvanian, "where alone it can be found, in a middle fortune—as far below wealth and splendor as it is above want."[16]

Without question, the majority of improving farmers held a fortune somewhere above middling, including merchant-squires of great wealth who dabbled as members of the more urban agricultural societies, together with the substantial planters who managed hundreds and even thousands of acres. Those who incorporated restorative methods almost always lived close enough to market towns to turn surplus into cash. Some raised the lucrative fibers—cotton and merino wool—not for homespun but for sale. And in a political universe split into two camps, they tended to vote for the party representing tariffs over the free market, the public survey over squatting, national development through internal improvements (like canals) over small government—the Whigs. No improver near great markets, certainly none among the planters, would have been mistaken for a *yeoman*. In an elastic definition offered by the historian Allan Kulikoff, yeoman farmers "understood land as a means to sustain themselves and their families, not to accumulate capital, even if they acquired substantial wealth and capital." The colonial yeoman family held a collective identity as part of a covey of linked households and sought to advance its material comfort without dependence on markets. Its insistence on noncommercial networks of barter represented an ethic above profit that may be its most defining characteristic.[17] The busy, commercially interested people who attended to dunghills in their barns look like the dire opposites of this standard.

Yet this cannot be the last word. As the isolated poles of rural class, *yeoman* and *capitalist* obscure more than they reveal and should be understood as points bracing a spectrum of wealth and identity. Improving farmers lived well within the rising market society and never opposed it; they used a manure-centered cultivation for commercial ends; yet they also expressed various forms of agrarian idealism. Some insisted that there need be no contradiction between the latest methods and log-cabin simplicity. Consider the posture of the *Plough Boy*, the journal of the Board of Agriculture of the State of New York and among the first of the rural newspapers. The editor, Henry Homespun, Jr. (a.k.a. Solomon Southwick), dedicated its pages to all those "who deprecate the moral laxity, the false pride, the dissipation and extravagance, into which thirty years of flattering but *fickle* prosperity have plunged this infant Republic." He championed the self-sufficient "homespun" family: "It was that family which first raised the standard of Independence in the revolution of 1776; and which never deserted that standard, until great-Britain was taught, that this people would not only make their own 'hob-nails,' but their own laws and government."[18] This from the same paper that announced every new breed, weed, and mechanical device and recommended the same intensive husbandry as the merchant-dominated societies of Boston and Philadelphia. No dire opposites here: Southwick never denounced the money earned from surplus crops, only the conspicuous way some people spent it.

All of this bespeaks the complexity of economy and identity in a world that could no longer be divvied up between mercantilism and Physiocracy, between Hamilton and Jefferson. The historian Joyce Appleby has given a name to the messy middle where most improving farmers dwelled. "Liberal republicanism" embraced economic independence without rejecting commercial opportunity. According to Appleby, in the states where wheat flourished between the 1790s and 1819, farmers developed "a core of common interests." The broad affluence of these years convinced Thomas Jefferson that economic independence and commercial agriculture could coexist peaceably. Increasingly, writers on the subject of political economy—merchants and farmers alike—began to regard Agriculture as the trunk of a great national tree, with Manufacturing and Commerce as its branches. This liberalized Jeffersonian vision describes the goals of most rural reformers, regardless of where they lived: "both democratic and capitalist, agrarian and commercial."[19]

Improvement also represented an ethic. It is not sufficient to treat improvement as capitalist ambition in agricultural form, nor to dismiss it as just another facet of the market revolution. Those who took up this "good system" of land use no matter what they grew or where they grew it, expressed a desire to endure, to persist, to cultivate as an expression of their stake in local society. The books, articles, and addresses that make up the published corpus of husbandry from the second decade of the nineteenth century into the 1840s asked the same question: What combination of crops and animals would most likely result in an auspicious fusion, feed the larder and the account book, make the agriculture of the United States the "great sources of its riches, and the right arm of its power," and stave off a hunger for fresh starts? Commercial opportunity explains none of the fervor behind improvement, but there is no reason to press the point too hard. Profit and persistence pulled the same cart. As soon as fertility fell below a certain threshold, harvests thinned and so did returns from market. For those incapable of spending the money to make manure, or disinclined to do so, the territories beckoned. "The wealthier the farmer," concludes Kulikoff, "the less often his family moved." Profit, at some minimum level, was integral to persistence. For the majority of farmers and planters who attempted it, improvement simply offered a way to live well enough without having to emigrate and a means of keeping land in good order before it passed to the next generation. Permanence of society, landscape, and home was the paramount value of improvement.[20]

Permanence was not the paramount value in the culture at large. During the period of its brief ascent, improvement stood in opposition to the most astonishing period of Indian dispossession and white settlement yet seen in North America. Thoughtful farmers understood that much more than soil is brought into question by waste and overexposure to wind and rain: the prospects of a land-breaking people. The deterioration of soil is the inescapable injury of agriculture to the environment, so its severity is a sign of the fealty or failings of any people who husband land. Signs of failing seemed to be everywhere, according to European travelers and improving American farmers. For them, the richness of well-managed tilth became a standard against which civilization in the United States could be judged.

LAYING WASTE

Americans in the old states saw the bones of the earth after the Revolution. More than a century before Kansas and Oklahoma became icons of ecological debacle, exhausted and eroding land north and south captured the concern of rural thinkers. Humans became parasites of soil. A symbiotic relationship would be the best that could be hoped for, but American farmers tended to drain the body that hosted them, and the vector always pointed to fresh blood in the West. That soils had little to give before they finally gave out became a familiar warning among cultivators who began to demand a better system. Wealth, they said, could be attained not by laying waste but only by that degree of investment in place that Americans found so difficult. The problem turned on a question: Could farmers learn to prosper by generating fertility in one place, even while knowing that numberless acres lay beyond the mountains? Even if waste followed the most avowed logic of self-interest, could it be rejected for another logic? It had been centuries since the farmers of Europe had considered that problem, and many of the first people to record the devastation of American farmland came from the island nation of Great Britain. In the comparison between landscape in the Old World and in the old states many Americans came to understand the bad economy of their agriculture. A tour through the eyes of foreign and domestic observers between the 1790s and the 1830s reveals that although some regions reported full crops and healthy lands, others did not. Though settled in many places for well over a century by 1800, the Atlantic seaboard rarely attained the high degree of cultivation visitors expected from an established country. On the contrary, certain districts seemed badly tended, hard run. Among the first to incriminate American farmers was an English farmer and esquire named William Strickland, who visited the United States in 1795.[21]

Strickland described a country ravaged by its farmers.[22] All the proof he needed that tillage in the United States languished he found in Virginia, which he called unhealthy and "much worn out." He called the red soils at the eastern foot of the Blue Ridge Mountains good land but for the people who owned it: "A richer district by nature there cannot be . . . but, like whatever on this continent has been long cultivated, [these lands] are nearly exhausted." Virginians did everything wrong. They trod their grain in the field to clean it but ended up making it filthy with dust, so the flour came out poor. The wild garlic they failed to weed out of their wheat gave "a most offensive taste to the flour." Oats, a mainstay of English farming, did not appear in many American farms, he stated, "and are every where bad, but those of this state, [are] of the worst possible quality." His deepest scorn went to American crops he hardly knew but whose damaging effects seemed to be everywhere obvious: "Tobacco and maize, which heretofore have been the curse of the slaves, are now, with the slaves, allowed by all men in Virginia, to have been the ruin of themselves and their country: the almost total want of capital, among this description of people, forbids all improvement on a great scale." Strickland's tirade could not be contained: "The land owners in this state are, with few exceptions, in low circumstances; the inferior rank of them wretched in the extreme. . . . Land in America affords little pleasure or profit and appears in a progress of continually affording less. . . . Virginia is in rapid decline."[23]

British travelers provided some of the most piercing accounts of American occupancy, but their commentary came with stern assumptions about landscape. They did not always make a distinction between farms well planted and well groomed, and along with their American disciples they demanded neatness and clean till in every instance. Strickland did not come from a country of smoky forest clearings, and he approached the use of resources in a very different way from most American farmers—as an accountant of sorts. He said that cultivation in the New World proceeded as though nature could be disposed of without consequence when in fact there would be a bill to pay. He added up the "costs" of land abuse and factored them into the money earned by crops at market, concluding that profits from the production and sale of grain are "certainly not nett," because "wheat and maize must pay for their neglected waste, and also for the worn out old-field, which produces little or nothing." He performed his own off-the-balance-sheet audit, reckoning the value of the harvest along with the productive capacity of the land. Well ahead of his time in this sense, he tallied the gross product and compared it with the "external" damage it caused. "Should this deduction be allowed, little profit can be found in the present mode of agriculture of this country, and I apprehend it to be a fact, that it affords a bare subsistence." Americans failed to internalize the cost of ecological damage even when it resulted in smaller and poorer harvests, and rather than make good what they had begun, they could join the "culprits" and other "outcasts" who ran away from the bad balance to the frontier.

Cultivators in the United States had more at stake than English tourists, and they also responded to the decline of farmland. After the Revolution the sight of a countryside increasingly distant from hardwood openings and gentle savannas began to grate on the sensibilities of farmers and travelers. Highbrow planters of an earlier generation, like John Beale Bordley of Maryland and George Washington of Virginia, complained in much the same vein as Strickland. Bordley compared the farms near his own with those of English proprietors, and the exercise sent him into a harangue. The tedious cycle of maize, wheat, and a year of rest with no cover crops resulted in lower and lower yields; and when "resting," the land was "all the while labouring under Oppression from exhausting, binding Weeds and Rubbish,—and the Hoof that beats it to a dead Closeness."[24] Washington said the same of his neighborhood: "The system of agriculture (if the epithet of system can be applied to it,) which is in use in this part of the United States, is as unproductive to the practitioners as it is ruinous to the land-holders. Yet it is pertinaciously adhered to. Our lands . . . were originally very good; but use, and abuse, have made them quite otherwise."[25]

Depending on the crops in rotation and the duration of the fallow, this "three-crop system" was either a modest way of restoring land or a rapid and profligate way of wrecking it. After wheat, grumbled Washington, "the ground is respited (except from weeds, and every trash that can contribute to its foulness) for about eighteen months; and so on, alternately, without any dressing, till the land is exhausted; when it is turned out, without being sown with grass-seeds, or reeds, or any method taken to restore it; and another piece is ruined in the same manner." Not English travelers but southern planters coined the phrase "land killing." George W. Jeffreys of North Carolina cried, "Our present, is a land-killing system," which, without reform, would "ultimately issue in want, misery and depopulation. . . . There is hardly a farm in the state of Virginia or North Carolina, but what exhibits the effects of this exhausting rotation of crops, in its galled and worn-out appearance."[26] In most cases, the problem had to do not with any inherent feebleness of soil but with the easy availability of cheap acres on the other side the Appalachian Mountains, which made any attempt at restoration seem like too much money and trouble. The flip side of the three-crop system could be found on any map of the United States: "Fresh lands of great fertility . . . at very low nominal prices, [have] greatly contributed to accelerate among our land killers, the exhaustion of our soil."[27]

Land killing followed a logic impossible to ignore. The reason American farms often looked so scruffy and downtrodden had to do with one of the most important differences between North America and Europe. One continent had an unimaginable extent of unclaimed land and the other did not. One continent had a severe shortage of available labor, especially in thinly settled outskirts, while the other had concentrated populations of landless people. The colossal imbalance between labor and land in the New World became a catalyst for mechanical innovation and a rationale for waste, because freeholders frequently moved to primary forest or grasslands rather than invest the wages necessary to improve acreage they already owned. George Washington said the same in a letter to the English experimenter Arthur Young: "The aim of the farmers in this country (if they be called farmers) is, not to make the most they can from the land, which is, or has been cheap, but the most of the labor, which is dear;

the consequence of which has been, much ground has been *scratched* over and none cultivated or improved as it ought to have been."[28] "Skim and scratch" described a casual cultivation, conducted with the least possible effort, intended to take what was easily available and leave the rest.

The pull of abundance even had the power to naturalize English farmers reared on restorative husbandry and turn them into American profligates. Said one who had seen this conversion in 1819, "I have lost my patience." Superior farmers arrived in the country "with a firm resolve to pursue their own plan of judgment," only to descend "into the slovenliness and absurd customs of the Americans." These once thoughtful husbandmen ended up sowing "the same kind of grain, on the same piece of land, seven or eight, or more years successively." In good American fashion they cared only for convenience "and the saving of labour." The American "seldom or never looks forward to the future and progressive improvement of his land; he uses it as asses are used in this country, worked while they have a spark of life in them, without one care about their support or preservation."[29]

If farmers had put their assumptions into words, they might have said this: the best way to use expensive labor is to put it to work on the most fertile land possible. Unbroken soils, free for the taking, "finance" maximum yield for minimum effort. Boundless expanses within the boundaries of the United States all the way to the Mississippi River made emigration more compelling than spending the money, learning the methods, and expending the labor necessary to create fertility. Furthermore, freedom from these costs allowed the great majority to own productive farms. Waste was democratic. Conserving land could be an expensive undertaking. Skimming allowed farmers to maintain yields with the smallest possible investment. Consider the political consequences of the landed endowment. Voting rights were pinned to landownership in most states until the 1820s. The Republic as a nation of freeholders—incorruptible so long as they held material independence from any merchant or monarch—found expression in the Declaration of Independence as "the pursuit of happiness." American equality, in some sense, depended on the resources that made every farmer the equal of every other and richer than any European peasant or wageworker. The entire republican project, inasmuch as it assumed that upward growth in population would force the outward geographical extent of the United States, was predicated on the waste of land.

No firm line separated field and forest in the continuum of land use. Every farm in the humid East began with clearing, so the squander of soil entailed a sustained assault against the woods. William Darby, who toured the North in 1818, recorded the consequences of fast turnover in the rural landscape: "The decrease of timber for building, fuel, &c. is already a great inconvenience, and is every day becoming more serious. . . . It may be said, that the tenures by which real property is held in this country . . . present an obstacle to any plans of permanent improvement."[30] Not only did farmers exhaust what they tilled, but they tilled too much and too lightly. "It is the custom with farmers to sow or cultivate a much greater quantity of land than they can properly manage; a consequence of which is that a great deal of good land is thrown away; producing about one third or one fourth of what it would if properly manured and attended to." The idea became fundamental to rural reform and ties that movement to conservationist thought. Farmers capable of making manure and applying it to fields in an intensive system plowed half or a third as many acres as the typical farmer and enjoyed a cascade of benefits: a more productive use of labor, a greater quantity of grain and grass harvested, and a surplus of land that could then be reappointed for the propagation of timber.[31]

The importance of all these criticisms to a general philosophy of improvement cannot be understated—they formed its core. Maintaining the fertility of soils and a balance between plowland and woodland served a particular conception of society. There will be much more to say about fallow, fodder, farms, and plantations in chapters to come. For now come away with this: these examples of agronomic reproach summoned a form of cultivation that was new and radical in North America. Controversy over the three-crop system only increased after 1800, especially in the South, where a former President of the United States also noticed the troubling decline of cultivated spaces.

One touchstone for the body of thought that began to emerge after the War of 1812 came from James Madison, retired from the presidency and in May 1818 presiding over his local agricultural society in Albemarle County, Virginia. His address appeared on the front page of every major rural newspaper and in the published memoirs and proceedings of agricultural societies in every region of the country. Farmers never simply altered their private property, said Madison; they tampered with environments and became implicated in the rest of nature. Farmers who did not steward their plants, animals, and nutrients lacked the longheadedness, the sense of future, required to build a republican nation. Thinking farmers like Madison came forth as the only people to caution that expansion did not necessarily create wealth or signify progress.

Madison began his lecture by endorsing Thomas Robert Malthus—the doomsaying parson who predicted that population must eventually outpace the supply of food. Though it did not seem to Madison that a "determinate limit presented itself to the increase of food, and to a population commensurate with it," a society could undo itself in other similar ways. The infinite extension of cultivated land would drastically reduce the diversity of organisms that made life possible and pleasant. Madison came up with an astonishing fusion for the time: "We can scarcely be warranted in supposing that all the productive powers of its surface can be made subservient to the use of man, . . . that all the elements and combinations of elements in the earth, the atmosphere and the water . . . could be withdrawn from that general destination, and appropriated to the exclusive support and increase of the human part of the creation."[32] Not only would humans consume themselves out of existence if they kept going, but they had no right to appropriate the entire creation for themselves.

So Madison counted up the creation and set it on a scale next to the needs of humankind, searching for the balance: thirty or forty thousand kinds of plants, six or seven hundred kinds of birds, three or four hundred quadrupeds, a thousand species of fish, and more reptiles and insects than anyone could count. A hungry humankind would eventually challenge this "profusion and multiplicity of beings" for the resources of land and replace them "with the few grains and grasses, the few herbs and roots, and the few fowls and quadrupeds, which make up the short list adapted to the wants of man." Madison may have discovered genetic erosion, and he used the concept to pound a hole in the hull of progress: "It is difficult to believe that it lies with him, so to re-model the work of nature . . . by a destruction, not only of individuals, but of entire species." Such an expansion would amount to nothing less than a "multiplication of the human race, at the expense of the rest of the organized creation." Madison's sense of interconnectedness extended to the fundamental parts of nature, and he seems to have believed that a ratio between plants and animals was responsible for life on Earth: "The relation of the animal part, and the vegetable part of the creation to each other, through the medium of the atmosphere, comes in aid of the reflection suggested by the general relation between the atmosphere and both." If either class should ever decrease too sharply, the atmosphere would be exhausted, the breath of life would cease, and the remaining species would not survive.

Here was a generous and voluminous awareness of nature and human needs, folded into a protean Gaia theory in which the biosphere heaved. It also asserted limits to wealth for any people who derived their livelihood from the ground, and this is where Madison was going. With nature so limited and space for people and other creatures in so delicate a balance, the critical occupation of humans in a populated world must be conducted with great care. Agriculture, the literal mucky methods people used to make food, could cause unimaginable harm. So its perfection would have an effect far beyond the furrow slice. The only safeguard of a republic, after all, is the virtue and good conduct of its citizens. So the conduct of citizens in the care of soils creates the nation or breaks it. In his conclusion to the essay Madison turns to "errors of husbandry." They compose a kind of Constitution with Articles, and though the thoughts of a venerated citizen, they are important for being entirely typical of countless tracts and addresses on the new conception of land use:

1. "*Any system . . . or want of system, which tends to make a rich farm poor, or does not tend to make a poor farm rich, cannot be good. . . . The profit, where there is any, will not balance the loss of intrinsic value sustained by the land.*" This is the First Principle guiding all others: there is an intrinsic value of land that is within the control of all tillers either to realize or to squander.
2. "*The evil of pressing too hard on the land has also been much increased by the bad mode of ploughing it. Shallow ploughing, and ploughing up and down hilly land have, by exposing the loosened soil to be carried off by rains, hastened more than any thing else, the waste of its fertility.*" Madison demonstrated a striking ability that we will see again and again among improvers. He comprehended the large-scale consequences of small-scale techniques. Even more piercing to the ears of Virginia planters, Madison admonished them to concentrate their slave labor on fewer acres as a way of conserving forested land. It was advice few of them accepted.
3. "*The neglect of manures is another error.*" There is so much to say about manure that it will take a section to lay it out. Madison pointed to the key ecological factor in all perpetual cultivation—the ability of farmers to return the nutrients they take. He said more than that. The neglect of manure represented a collision course between nature and the market, because when the entire product of a farm goes to town, never to be consumed where it grew, land suffers a debit. In China land is never turned out to fallow, Madison averred, because "an industrious use is made of every fertilizing particle [from both animals and humans], that can contribute towards replacing what has been drawn from it." He elegantly equated the internal cycle of a subsistence farm to that of a forest, where "the annual exuvae of the

> trees and plants, replace the fertility of which they deprive the earth." Thoughtless consumption on a gigantic scale, causing nutrients to move in a one-way direction away from fields, may be the only force of humanity really to be feared: "With so many consumers of the fertility of the earth, and so little attention to the means of repairing their ravages, no one can be surprised at the impoverished face of the country." Madison said that when they exchanged fertility for dollars, planters sold the stability of rural society.

Judicious and precise, the address drew a blueprint for how to cultivate in the same place without the need for emigration. It is the voice of a former Federalist who worried about a delicate natural order falling out and a lack of elite leadership to guide the populace. It is not the voice of a Jeffersonian, for whom nature never came into conflict with republican government, for whom the words "finite amount of land" could never be spoken or thought. Madison, of course, used both voices at different times.[33] The address of 1818 was also important for what it was not. It lacks sectionalism and politics. In the mood of nationalism that followed the War of 1812, the Federalist Party crumbled, and parties themselves declined in importance for a time. The appearance of a single national interest lived only briefly while the thorny implications of tariffs and the second Bank of the United States lay unexamined. So in 1818 Madison talked about farmers, not planters, labor, not slavery, and addressed the whole nation on the errors of its agriculture without regard to party or section, as if those things did not exist.

Seven months later the earth shook. Madison must have felt tremors, as did farmers all over the United States, who began to clamor about negligent practice, seemingly conducted by the same invisible maestro. What hit the ground in 1819 did more damage than drunken redcoats with torches. It hit hardest as far away from Virginia as Ohio and ruined livelihoods for the next ten years. The Panic of 1819 set off the first great depression in the United States and caused a panic of a different kind among farmers and planters, who worried that they had lost the value of their land. It is impossible to separate larger economic and political events from changes in the management of thousands of individual farms, and only by seeing the panic as people at the time saw it will the practices themselves have meaning. A financial panic could be a soil panic, and vice versa.

PANIC

The explosive postwar economy, rife with debt and speculation, finally went down in late 1818, and the United States entered its first great depression. The shock left no fragment of the nation untouched. And because farmers made the great majority of the commodities and carried the majority of the nation's private debt, their response to the crisis is especially important. Financial events can change the countryside in regions far away from banks, and the panic forced farmers to make difficult decisions in circumstances they had never seen before. Henry Clay described the effects of the depression later in the 1820s: "It is indicated by the diminished exports of native produce; . . . by our diminished commerce; by successive unthreshed crops of grain, perishing in our barns and barn yards for the want of a market; . . . and, above all, by the low and depressed state of the value of almost every description of the whole mass of the property of the nation."[34] The diagnosis of political economist Mathew Carey hung like smoke over the landscape and captured the fear of the moment: forced sales, bankruptcies, scarce money for borrowing, suspension of large manufacturing, overflowing prisons, countless lawsuits, defaulting families in irons to pay for day-to-day expenses—all added up to a society in economic and social arrears. One speaker at a cattle show told his audience, "I am an alarmist": "Last year we talked of the difficulties of paying for our lands; this year the question is how to exist . . . families naked—children freezing in the winter storm. . . . As a people we are growing poor." *Niles' Register* estimated that fifty thousand people languished without employment in the three largest eastern cities, and by another estimate half a million went unemployed throughout the country. Sympathetic stories of desperate theft filled the papers, and the convicted accepted incarceration with gratitude.[35]

The panic brought into question the profitability (or in economic terms, the rent) that a farmer could expect from quality farmland. During a brief interval, and in certain locations, improvement became a crusade in which concerned planters and farmers articulated practices intended to keep themselves in business and in place during the bust. The reasons for their fear had to do with the panic itself.

Cotton prices, a benchmark indicator of the early national economy, soared between 1815 and 1818, reaching 32 cents a pound. The prices for many other commodities also reflected spectacular growth and runaway demand as the United States concluded its second war with Great Britain. But when European production picked up again after the peace, English textiles began to re-enter American markets. Worse, high prices for American cotton sent British manufacturers looking for a cheaper source, which they found in India. By the end of 1818 the price of cotton at Liverpool had begun to flutter before it fell to 26 cents. American markets reacted with one sharp drop followed by another, so that although the price seemed to bottom out in 1821 at 13.5 cents, after a rally of sorts in 1826 it crashed again, landing at 11 cents.[36] The second Bank of the United States once fueled the boom in trade and manufacturing with expansionary credit; it now brought on the bust by demanding its major debts. Autumn of 1818 marked the scheduled repayment for the Louisiana Purchase, and the bank had been in a deflationary mode since summertime, calling in notes from branch banks, holding on to its gold.[37] Sinking cotton and shrinking credit threw a one-two punch felt from Atlanta to Cincinnati. Businesses failed; banks failed. The economy, it was revealed, did not stand on solid prairie but quivered on stilts over a swamp.

Planters had purchased additional land to rake in the money during the flush times, and many northern farmers had gone into debt with merchants, thinking that high prices would always be there to help them pay off the cost of furnishings and other goods. The quality of production hardly mattered as long as upland cotton went for 32 cents. Before the crash, planters believed that they could make money with the least imaginable work and attention. Said one from South Carolina, looking back: "In the halcyon days of high prices and prosperity, when mother earth was made to render her treasure at any and every cost, it was thought sufficient to constitute a planter to be the owner of lands and slaves," while the real management went to an overseer with no interest or stake in the plantation.[38] Now, with prices low and lenders asking for payment, slovenly fields no longer yielded at profitable levels. A dismal reckoning took over. The market tied the value of land to the value of crops, so when prices fell, so did any accounting of the invested wealth of farmers. To a family mostly content with producing for themselves and with no thought of moving, that fact might not have mattered very much, but for anyone with heavy loans against their holdings the panic was much more serious. When the work of slaves resulted in lower returns, their value also suffered. Powerless over prices, planters and farmers could either escape to the frontier or look inward to the unrealized productive capacity of their land.[39]

For rural people the panic began before it began. Some of the most pointed and frightened declarations came the year before, even with cotton prices high. The deflation that began in the summer of 1818 had everything to do with this anxiety, but something else scared planters even more than the banks. Tench Coxe, protégé of Alexander Hamilton, published an appendix to a kind of policy brief he had first issued in 1817 on the subject of the international cotton supply. What Coxe had learned in the meantime about the extent of the trade between Britain and its colony on the subcontinent made him shudder and rush to press. Freight from India to Britain ran cheaper than it once had—bad news, but only the beginning. India enjoyed a longer and hotter growing season, and the extent of its cotton land was greater than the entire Louisiana Purchase. "*The ruin of the cotton market at home,* as well as abroad, would occasion much of that great portion of population to turn to the cultivation of grain, cattle, tobacco, rice, and sugar, so as seriously to affect their prices, and injure our general planters and farmers. . . . No fact or prospect threatens our agriculture with *so much substantial evil,* as the rivalship of our cotton wool by foreign cotton wool."[40] To avoid being pauperized, planters, urged Coxe, should put in other crops, at least during 1819. Even news of high prices carried dark undertones the year leading up to the panic. *Niles' Register* remarked that the enormous value of cotton "is, to our southern brethren, far more productive of wealth than the mines of Mexico and Peru," but the article ended with this: "The British still talk much about supplying themselves with cotton from India—last year they received 30,000 bales, and they say they expect 150,000 bales in 1818."[41] Coxe said that India had 220,000 bales packed and ready to ship.

This is the early intelligence that may have prompted James Madison to write his address, and it animated planters and farmers in almost every other state to call for wiser cultivation. Farmers also felt the lurch and hurl of the manufacturing economy, now up again, now down, with surges in imports countered by tariffs to slow the pace of commerce. Though the panic seemed to affirm agriculture as the only sure fountainhead of the Republic in the eyes of some, it also placed pressure on farmers to make up the difference in national wealth. Another effect of the crisis (mostly in the North) was to drive entrepreneurs back into agriculture, thus, according to Mathew Carey, "converting customers into rivals, increasing production, and diminishing the home market."[42] Many of the people who set up factories had recently been farmers, and some had produced the very raw material that they now milled or spun into finished products. The enterprising might wear one hat or another, depending on the advantage of the moment. A writer known only as The Speculator recommended this as a policy of survival: "Every man who reads this, who is gaining nothing or going behind hand, should turn his attention to agriculture."[43] He was talking to the merchant without customers, the physician without patients, the lawyer without clients. This "backward shift," as it was called by those who saw it as a falling back from the march of civilization, had begun in 1815 and continued all through the depression of the twenties. Madison mentioned it too, observing that "the manufacturer readily exchanges the loom for the plough, in opposition often, to his own interest."

For those who refused the lure of fresh fertility, all this chaos made them blink at the sight of land in cultivation. During the expansion that lasted from the 1770s through the Monroe administration, when wheat was high from Virginia to New York and farmers shipped it across the Atlantic, improved land could be bought and sold with confidence and represented nearly all the money that a family would ever see. The bust weakened that confidence and forced farmers into another kind of calculation. If the little capital that farmers invested in arable acres became stagnant, then their only refuge was to make the most of the ground they owned. Not land as paper affluence but land as dirt and plants—that is the change that the panic brought. In an instant, the ways that farmers created food and fiber had enormous implications for their own wealth and the prosperity of the United States. Those farmers who, because of political interest, economic obligation, or moral persuasion, refused to consider a life in the West lashed out at waste as they felt the discomfort of ecological limits.

Gathering in 1818, the members of the Society of Virginia for Promoting Agriculture considered their plantations as links in a chain extending from soil to practice to politics. The metal had worn thin:

> A soil originally fertile, has been rarely improved; and has, in many places, been reduced to such a state of sterility as scarcely to compensate the expense of cultivation. Taking possession of an immense wilderness, covered with thick forest, our ancestors were compelled to employ immense labour in clearing it. For a long time their utmost industry scarcely enabled them to open a sufficient quantity of ground to furnish subsistence for their families. Continual cultivation was produced by necessity, and exhaustion was the unavoidable consequence. New lands invited and rewarded the labourer; and cutting down and wearing out, became habitual. The effect belonged to the cause, and flowed naturally from it. But the cause exists no longer. . . . That necessity which originally impelled us to cut down, now impels us, with a force no less urgent, to restore the fertility of which our soil has been deprived.[44]

The effect belonged to the cause. Landwash and squander became an economic problem when the price of cotton faltered, transforming once-adequate low yields into a sterility that scarcely compensated the expense of cultivation. Sensing the coming decline months before the banks did, these planters declared themselves against cutting down and in favor of restoration. The *Memoirs* of the society continue from this preface to papers on various subjects relevant to the dire conditions of land in Virginia: the cultivation of Indian corn, artificial grasses, the Hessian fly, the rotation of crops, plaster of Paris, defects in agriculture.

James Mercer Garnett of Virginia claimed that worn-out land returned to wilderness. He reported seeing farms "now almost entirely overspread with vast and gloomy thickets," deer "more abundant in several of these counties" than when they were occupied by Indians, wolves haunting the night as they once did, pine and cedar sprung like weeds in the footprints of long-gone planters—this was Virginia. Garnett did not shrink from the largest possible implications of the bad system—the slackening of civilization in his state, its accelerated depopulation, its widening poverty. In his own study, the large number of acres recently offered for sale indicated an agricultural crisis. Planters sold for many reasons, he admitted, "yet I believe it may be assumed as a general fact, that very few, not enough to form an exception worth calculating, would sell, unless they found agriculture *here*, a losing business." Garnett examined two newspapers, where he discovered 21,773 acres advertised. By extending that quantity to the twenty other newspapers published throughout the state, he calculated the number of acres offered in 1818 at close to 500,000.[45]

Northerners feared the depression no less. Samuel H. Black, speaking before the Agricultural Society of New Castle County, Delaware, described destruction, denial, and emigration:

> Whatever diversity of opinion may prevail on the subject of farming and on the value of land, that both at present amongst us, are in the most melancholy and wretched state of depression, will, I think, be readily admitted by all; that crops of every kind, have of late years, almost totally failed; and that, to the laborious husbandman, scarce a single hope seems left at which he can grasp to encourage him to new efforts . . . ; this

> gloomy truth is felt, and admitted in every sphere of life, from the pauper in the poor-house, to the most wealthy man in the state.

The owners of land attributed the problem to "every other, than the true cause—the seasons, the climate, the Hessian-fly, the stunt, the louse, the grub, the clay and the sand, with an endless variety of other imaginary evils. . . . The tiller of the ground has perhaps hardly dreamed, that in execrating the cause of his ill success, he but calls for the vengeance of Heaven on his own head. All these plagues are but *symptoms*, and not as he may have supposed, the *cause* of his misfortune."[46] Robinson Crusoe used up all the wood on his island, Black argued, but no one blames the wood or the fire or the sultry air for his loss of comfort. Crusoe's island presented him with obvious limits, while most American husbandmen saw the next county or the next state.

On the brink of depression, the Philadelphia Society for Promoting Agriculture committed itself to a mutuality of agriculture and commerce. Founded in 1785 and the most respected organization of its kind in the country, the PSPA counted a membership both active and honorary that included Nicholas Biddle, president of the second Bank of the United States; Mathew Carey, a political economist and advocate of manufacturing; Senator Henry Clay of Kentucky; Tench Coxe; Noah Webster of Connecticut; Josiah Quincy, a merchant and gentleman farmer from Boston; and James Pemberton Morris, a farmer from Bucks County, Pennsylvania (whom the reader will meet again).[47] Its president announced in November 1818 that all the cash and hard goods changing hands in the merchant houses, evidence of "the rapid accumulation of wealth from a prosperous state of foreign commerce," had the effect of taking public attention away from agriculture, which had been left alone "to struggle against every obstacle." As judges and financiers saw it, the hardships and embarrassments of the depression "prove how necessary a prosperous state of Agriculture is to the success of Trade." After all, "soil is the basis of national wealth, and its cultivation the only permanent source from which its prosperity can be derived."[48]

Farmers would never join the humming world of commerce and manufacturing as equals as long as they resisted a rethinking of their fundamental practices. Like the Virginians, Frederick Butler of New England argued that farmers needed to learn a new system altogether, one that would be a studied rejection of what they had been taught by the generation before them: "Let us remember, that this system of husbandry which I have laid down, would have no more, and no better, applied to our fathers, than to the present inhabitants of the fertile wilds of Illinois." In both cases settlers lived in the illusion of limitless gain. "The more corn they could plant, and the more wheat they could sow, the more profits they derived." That way of thinking placed rural production on a collision course with capitalist opportunity because it failed to account for scarcity. The countryside of the seaboard states had grown to maturity since the Revolutionary generation, and farming would have to change too: "What in them, under that rich state of soil, was good farming, has become to us, under the exhausted state of the soil, bad farming." Some anticipated that the depression would force a new paradigm. As one textbook for schools announced in 1824, "The time will come, and indeed in many places now is, when the land, repeatedly wounded by the plough-share, and exhausted of its riches, shall be too weak, of itself, to make plants grow with their former luxuriance. This may be called THE ERA OF SYSTEMATIC AGRICULTURE."[49]

With so much converging on it, the furrow slice became a central location where Americans contemplated nature and economy in the early nineteenth century. Uninspired by distant river valleys, disdaining those statesmen who urged Americans to break the continent to the plow, an important minority of farmers and planters decided to dig in, preferring to rethink agriculture rather than remake their world on the frontier.[50]

At the center of these concerns stood a pile of dung. Unattractive and strange to the uninitiated but a stern monument to those who knew its ways, dung held great power. If improvement and all that it stood for had a single symbol, it was this—the steaming excrement that completed a circle in the land large enough to enclose the riches of rural life, strong enough to make the farm equal in strength to the changes taking place in the upstart sectors of the post-1812 economy. If the farm would stand with manufacturing or against it, one thing was certain: it would have to make manure. And to those who claimed that the glory of the United States lay in the West and that the unceasing exploitation of soil would result in a prosperous nation, the dunghill argued otherwise. Onto this single hot and aromatic structure improvers heaped all their hopes and goals: a permanent rural society, the leadership of responsible elites, a countryside distinguished for its beauty and neatness, the application of reason to artifice, and various desires for integration or isolation from the wider world—all of it seemed possible when dung got mixed up with soil. The dunghill seemed to offer a way out of the paradox of a declining environment that would provide the raw material for an economic revival. At the moment they realized that agriculture had resulted in widespread degradation, farmers all over Atlantic America came to believe that the same soil could bear a great deal of economic and political weight.

DUNGHILL DOCTRINES

Emerging from the depression in states up and down the Atlantic seaboard was an American response to "land killing" that called for the union of cattle and grass in a synergy that created dung. Not the dung of the riverside and the woodlot, but dung that could be saved, composted, fermented, and applied to specific fields at specific times of year. Many readers from city or suburb, like me, may judge dung on its more obvious qualities. It smells bad and harbors bacteria. Ordinarily it should not be left lying around on the ground. But farmers of the past knew it better. They knew its texture, its tendencies when fresh or rotting, how it piled, how it spread, how to collect its "juices" in the barnyard so that nothing would be lost. They knew the stuff for what it was—the partially digested wealth of their farms in a form that could be hauled to fields and made to turn another crop. The Virginia farmer who allowed his animals to graze along the branch stream or among the pines could never create manure; the Pennsylvania farmer who let the stuff pile up around the barn until it blocked the doors also let the fountain of fertility run into a ditch. Manure needed to be collected and cared for; it needed to be dug out from pens, layered with straw, covered from the wind and sun. For John Taylor, a Virginia planter and author of a collection of agrarian essays published as *Arator*, well-dunged soil was to its alternative what the pippin apple of the orchard was to the crab of the wilderness: tame, plenteous, domesticated—a victual that fattens animals and makes the country sweet.

It begins low in the gut of the animal, in brothy, vaporous chambers. Once course fodder from pasture or pen is taken in by lips and teeth, it is chewed, mixed with saliva, rolled into a bolus, and ejected into the anterior rumen, the first of four compartments in the marvelous ruminant stomach.[51] Billions of bacteria and protozoa inhabit the vatlike gut. There they ferment the green food and through their own digestion produce great domes of gas, which must be voided to prevent dangerous bloat in the host animal. Rumination is the regurgitation of the ingesta and the re-forming of the bolus for further chewing and re-swallowing. A period of rumination may last two hours. Chomped to fine particles, subjected to more fermentation, still blooming volumes of methane, the ingesta moves through the stomach: reticulum, omasum, abomasum. Most of the nutrients are absorbed in the small intestine before the undigested feed, now fecal matter, spills into the lower tract. Then the muscles get involved, and the bowels finally move, and then comes the exit. A circle in time and space completed, though the cow is unimpressed by it (to say the least).[52] The inside of a cow seems designed for the difficult task of extracting nutrition from a substance—cellulose—that does not easily give it up. After the entire process has reached its conclusion, the undigested leftover still contains a surprising amount of its original nutrients. It is as though the cow borrows fodder, then gives it back, only somewhat worn.

Dung that returns becomes manure, the name for any substance that augments the fertility of soils. *Manure* is both verb and noun—the word is used to refer to the process of applying a particular material for fertilization as well as to the material itself. When it falls out hot, call it dung; when it goes to rot in mounds of straw, call it manure. There is green manure (any crop plowed under) and mineral manure (lime or gypsum). Manure, unlike topdressings (fertilizers applied once crops are in the ground), is always applied before planting. And yet dung by any other name did not smell quite the same to nineteenth-century cultivators as the animal or "putrescent" manures. They included the dung from birds (also known as guano), town dung (from horses), night soil (from people), urine, and fish.[53] The domestic quadrupeds all provided farmers with fine feces, but none provided the quality or quantity of that which came from cattle. Farmers prized this stuff most of all, and for many writers manure and cow dung seemed to be the same. As the New York editor Jesse Buel put it, "Cattle and sheep make manure,—manure makes grain, grass and roots—these, in return, feed the family, and make meat, milk and wool;—and meat, milk and wool are virtually money, the great object of the farmer's ambition, and the reward of his labors. This is the farmer's magic chain, which, kept bright by use, is ever strong and sure; but if broken, or suffered to corrode by neglect, its power and efficacy are lost."[54] When people structured the farm to feed the furrow, they fed themselves for centuries.[55]

The payback was remarkable. Not only did cow dung contain all the nutrients that plants took from soils, it contained them at high levels and in a form that plants could easily put to use. The digested plant matter that comes out the hind end is almost identical in its chemical composition to what goes in the front. As a rule, notes the soil scientist Nyle Brady, 75 percent of the nitrogen, 80 percent of the phosphorus, and 90 percent of the potassium found in the feed are voided by the animals. Yet the manure itself is mostly water—between 60 and 85 percent—so it might contain only 2 percent by wet weight of each of the three key nutrients. Manure keeps on giving for years after its last application. Researchers in a famous English study dunged a test plot with thirty-four megagrams of farm manure for twenty years between 1852 and 1872, then stopped. The plot maintained a higher level of fertility for the next hundred years than ground never fertilized.[56]

The question for any farmer was how much to apply for best results. There was no adequate answer during the nineteenth century, since how much to apply depended on the expectations farmers had for the size of their yields and the nutrient composition of a load of dung depended greatly on the quality of feed that cows consumed to make it. At the average composition of 0.5 percent nitrogen, 0.25 percent phosphorus, and 0.5 percent potassium per unit of solid dung, it takes 2 metric tons (2.2 short tons) to equal just 100 kilograms of the supercharged factory-made fertilizer popular today. Farmers intent on using animal manure to duplicate the punch of synthetic mixes need eighteen to twenty tons per acre.[57] Richard Meinert estimates that cows of almost two centuries ago, with lower-quality feed and breeding than their descendants, would have voided about 120 pounds of dung and urine in a day. An annual range of between five and ten tons per acre would probably have satisfied improving farmers of the 1820s. The low end represents the absolute minimum to maintain soil organic matter; the high end equals the yearly production of half a cow. In other words, keeping ninety acres of arable safely in crops required the digestion of forty-five cows.[58]

John Lorain of Centre County, Pennsylvania, recorded a number of key statistics describing his own production of manure that allow for some of the best estimates of the needs and capacities of the 1820s farm. Lorain calculated the weight of roots, grasses, and cornstalks, added that to leaves from his woods, and had the whole mash "saturated well with the juices of the cattle yard, [to] form a great weight." The weight came to fifty-four tons, gladly consumed by his cattle to produce fourteen loads of manure, thirty-two cubic feet each, in an undefined period of time. A "load" in this case weighed 2,742 pounds, so this dung yield came to over nineteen tons—about what ten nineteenth-century cows could make in one month. The farm in Centre County enclosed 106 acres, of which Lorain used 20 for "convertible husbandry," or the periodic conversion of different parts of the same farm from fodder crops to grains to meadow. He kept 4 in clover or turnips, 4 in wheat or oats, and 12 in grasses. Divide the tons per year by 20 acres, and the result is 11.5 tons per acre per year—the digestive output of about half a cow per acre. Lorain would have emptied his pens once every two or three days because ten cows would have discharged three thousand pounds of dung in that time. Under this system, said Lorain, "the quality of manure exceeds credibility." He made four times as much money as a neighbor with the same number of acres who did not practice the good system.[59]

Improving farmers did not simply keep the stuff—they invested in it. Animals and grass may have been the raw materials, but alone they did not generate fertility. Manure required land enough for pasture, fields for "high feed" like turnips, buildings for keeping animals over the winter or all year and to store the pile inside, labor to take it from where the animals made it to where it needed to be applied. A proper dunghill is more than a reeking mass. Because dung contains so much water, the best advice recommended that it be set atop a broad floor of flat stones at an incline leading to a drain, so that the "juices" and "liquors" of the barnyard might be emptied into a well, then pumped as needed to the land. Those juices needed to be cared for when they flowed in abundance, so farmers needed to build reservoirs and wooded flumes to convey this very special runoff "for the purposes of irrigation."[60] The dung itself may be stored by first mixing it with straw to absorb the liquid. "Too often," stated an expert, the fluid "is suffered to drain away from the dunghill, so as to be entirely lost to the farm." Sometimes the hill was set over a layer of peat, marl, and chalk in a further attempt "to absorb those valuable juices that otherwise might be lost."[61]

The great divide in all these requirements, the one that drove a wedge between those who did and did not practice the new husbandry, came down to labor. How to remove three thousand pounds of dung from a barn twice a week? Farmers either found the hands to do the disagreeable work or lost sight of the sun as their barns filled up. Lorain told of people forced to burn their barns to the ground, a plan they called "more economical than encountering the labour of removing the manure," and it's not difficult to believe. The stuff then had to be shoveled into a cart and moved to fields near and far for an even spread. Those short of workers often put too much on convenient fields and gave little or none to more distant places. Turning it under added another task, but it guaranteed that the manure would become part of the topsoil without being carried away by wind or water. It all added up. Lorain spent $522 a year to pay a man and a boy to haul loads to fields and realized "that much labor, and very considerable expense, will be found necessary to restore the grounds to their original state of fertility."[62] A native of Fauquier County, Virginia, said the same: "To make poor land rich, it must be admitted, is at all times an expensive and difficult business—it furthermore is a work of time."[63]

Though anyone with the desire could plant a little grass and pen a few cattle, the people who operated a complete system raised manure to a top priority. Making manure, more than any other single practice, represented the intention and wherewithal of any farmer to remain on land presently cultivated.

None of this explains very much about the process of making dung in great quantities. To find out about that, we need to reach back to the source of the core practices. The people who thought most about increasing returns from a limited space during the nineteenth century were members of the British aristocracy and their tenants. They called it field-grass husbandry and also alternate, convertible, and up-and-down husbandry—all names for the same intensive integration of animals and arable fields. The best way to understand it is by what it replaced in the English landscape. In the old "open field" system, arable (plowed fields for grains and row crops) and pasture made more or less permanent patterns, with a fallow (year of rest) rotated through, much like the three-crop system of the South. Open fields depended on rotation as a way of replicating in a fixed space the practices that farmers and pastoralists once employed over vast regions. Where ancient tillers cultivated fields once every few years or took roots and other plants without formal cultivation, by about A.D. 800 Europeans had initiated a two-field system in which they fallowed for fertility. Where the keepers of flocks once traveled in seasonal migratory circuits, burning the ground in mountain meadows and forest understories to produce forage for their next visit, by about the same time European animals grazed on permanent pasture. The historian

Stephen Pyne describes this more sedentary pattern: rather than rotate the farm through the landscape, Old World farmers learned to rotate the landscape through the farm.[64]

The problem with the open-field system for the British gentry was that fallow very often fell short of replacing all that crops removed. Without the addition of manures or a cover of clover (which transfers atmospheric nitrogen to soils), a year of rest accomplished next to nothing and simply had the effect of wearing out tilth over a longer period of time than constant cropping. It takes much longer than a year for organic matter to accumulate and form a little humus (a name for the decaying organic matter on the surface of forest soils). In fact, depending on the environment where people tried them, systems of burning and deferral sometimes required a downtime of up to ten years. In a time and place of land abundance, long periods of dormancy could be endured by shifting cultivated spaces through an extensive landscape. In a time and place of land scarcity, the restoration of land needed to come from within the farmstead. Historians have long assumed that such a method first appeared during the seventeenth and eighteenth centuries, when the British gentry asserted rights over ancient common lands and finally "enclosed" them behind hedges by the blunt force of parliamentary acts. Having abolished all communal rights and obligations, having nullified local custom, they became the masters of great tracts, now squared and consolidated, where they created a landscape that reflected their aesthetic and financial interests. Their desire for profit and continuous cropping led the aristocracy to generate fertility. Here is the crux: without convenient wastes or open-field fallow to provide them with nutrients, landlords required a contained system. They learned to rotate the landscape through the farm with a greater intensity than had ever been attempted before.[65]

The only problem with this version of the story is that it gives the landlords much too much credit and places the Agricultural Revolution too squarely within the period of enclosure. Enclosure did hasten the new husbandry in some locations, but the innovations that built the "backbone of the Agricultural Revolution" first appeared in England in the 1560s and had emerged more or less in their final form by 1720. The landlords did not invent convertible husbandry; they adopted it.

Keep your eye on the grass as it goes up and down—plowed up for crops, then put back down in temporary pasture, called leys. Progressive farmers kept up to half of their acreage in some kind of pasture (permanent and temporary), with the rest of it in tilth. They converted their arable fields from one use to another in yearly rotations or courses like this: (1) wheat or barley the first year, (2) wheat or barley again the next, (3) clover or peas for hay and soil nitrogen, (4) wheat or barley, and (5) a fodder crop like turnips or back to grass. The return of nutrients in the form of manure resulted in larger crops of grain and fodder, which allowed for a greater number of animals to be kept on the same land, which created more manure—a positive feedback loop.[66] What did not go to market or to rent or to feeding the family passed through the guts of animals, so animals became the nexus of the farm or, in the words of John Sinclair, an agricultural theorist and contemporary of Arthur Young, "machines, for converting herbage, and other food for animals, into money." Convertible husbandry did not merely rotate functions within the same limited space; it maintained the dynamic balance between production and decomposition that is one of the crucial control points in any ecosystem. Most spectacular of all for the history of agriculture, the new husbandry eliminated the fallow and made every farm into a manure factory, and under this system the cycling of fertile nutrients became the central ecological function of agriculture. In a word, by conserving its own resources, the old farm replicated in a managed sphere the nature that humans did not manage.[67]

New intensity came from an old source. The alliance between animals and grasses was nothing new—not in the 1790s, not in the 790s. Until very recently, whenever humans wanted to shake up environments they struck a deal with one or another of the grasses. Paleolithic hunters knew that when they burned the forests and grasslands covering southern Europe, they encouraged the northern migration of the beasts they liked to eat. Perennial grasses thrived on soot because they reproduced through rootlike systems extending horizontally just underground (called rhizomes) that formed a web impervious to fire. But agriculture did not owe its foundation to bunchgrasses. It exploded from seeds. Wheat is not very impressive held up by the roots, but look at the other end, and here is a plant that invests its sun energy in sexual reproduction. Wheat will germinate in almost every extreme of climate and topography, from mountain meadows to desert heat, from the tip of Argentina to within two hundred miles of the Arctic Circle. It is now the most commonly cultivated plant on Earth. Once unified on the ancient Eurasian farm, the complex consisting of wheat, barley, cattle, and sheep moved along the boundaries of human settlement and European colonization.[68]

Convertible husbandry represented an intensification of the Eurasian complex, and it also emanated from the propagation of grasses. The grasses implicated in this revolution came up thick and ripe when planted in rich muck and tasted good to the bovine tooth. Chief among them was timothy (*Phleum pratense*), the unsurpassed hay-crop species, also known as meadow cat's tail and herd's-grass. Its fruiting head stands six inches off the stem in a long spike of dense flowers somewhat resembling the common cattail of marshlands but not at all like it. It grows well when combined with legumes like red clover, alfalfa, and peas. Timothy seeds weigh forty-five pounds to the bushel—a terrific yield of calories for little more invested than the labor of scattering it. Lucerne, white and red clover, bluegrass, tall meadow oat, sainfoin, burnet, trefoil, tare, and (on occasion) barley—these are the so-called artificial grasses native to England and Europe. They made a richer yield of dung by far than the "natural" grasses, like the wild oat and poverty grass farmers found poking up in the mud around hoofprints in the spring.[69] Nothing impressed a northern farmer as much as the sight of large and healthy cows grazing a clean and fine-seeded meadow, and the reciprocity between cattle and grasses is the essential biological relationship in this story.

Timothy leys came to North America as emissaries of a foreign power. The American tutorial conducted by British authors should not be underestimated in its importance, for the image of rural England—sod green and garden damp—deeply affected admirers in North America. The young Frederick Law Olmsted took his first breath of it in 1851: "The country—and such a country!—green, dripping, glistening, gorgeous! We stood dumb-stricken by its loveliness . . . homely old farm houses, quaint stables, and haystacks . . . the mild sun beaming through the watery atmosphere."[70] Look! a real Hereford cow, a real hawthorn hedge, a real English cart with a real ruddy-faced "smock-frocked" carter on board . . . real flowers! Olmsted sounded as if he had just discovered the original document from which all copies of refined landscape had been made. While Americans seldom said so outright, the mental pictures they held of rural comforts and land in good keeping came from British authorities. County reports in the published volumes of the Board of Agriculture made their way from London to the United States, where conscientious farmers read their detailed descriptions and assembled composite images of ideal landscapes. Cattle and rich grasses drove the greater aspirations of improvement toward a countryside worthy of being framed in the mind and on canvas.[71]

No English writer made a deeper footprint in this soil than Arthur Young. When George Washington retired to his seat, determined to give it a sense of order and appearance reflective of his own, he opened a long correspondence with Young. More a skilled and energetic publicist than a true innovator, Young took more credit for the new husbandry than he deserved. Yet the manner in which he seemed to scrape the mold off older ways of doing things appealed to certain American farmers looking to lend agriculture the same sophistication they associated with the professions. Young showed them a grassy, burgeoning, and patrician rural life, which remained in the imagination of educated farmers from Washington's time to Olmsted's.[72] Americans entered Young's world whenever they opened his many writings, especially his fits of worship at the great country halls. Young believed in the power of the landed class, and he made its success the object of all his effort. He supported the Corn Laws, which placed a tax on imported grain favorable to the rich; he despised the common lands and the wastes, urging their enclosure; and he enjoyed a mutual admiration with George III, known as "Farmer George" for his delight (at a clean kingly distance) in the new agriculture. There may be no better way to make this point than to follow Young during one of his diligent tours, and there may be no better place to visit with him than the famous Holkham, seat of Thomas William Coke, Earl of Leicester, one of the most important of the eighteenth-century improvers and inventor of the "Norfolk system," a successful local variation on convertible husbandry.

Even from the page, Young trembles at the thought of Holkham. His ascension from pedestrian along the public road to guest of the manor is related with rising anxiousness. The ascension begins with the slightest gesture of landscape, quiet enough to miss, "a few small clumps of trees, which just catch your attention, and give you warning of *an approach*." The approach opens to a triumphal arch, from which point it is a mile and a half to an obelisk. First pass the lodges of two porters, then start up a hill, see a number of cultivated fields on both sides of the road, reaching into the distance, finally arrive at the obelisk on the top of the hill, and there see eight views: lake, town, arch, fields, sea, church, planted hill, and the great house. Young finally arrives dizzy and disoriented and, after some difficulty finding the front door, enters "the inside of the house!" Corinthian pillars mark a marble passage in a space so large that "all sort of proportion is lost." He sees the salon and the dressing room and pauses to catalog some of the greatest paintings in Europe: Rubens's *Flight into Egypt*, Titian's *Venus*, one by Raphael, a number of pastoral landscapes of Poussin, and Claude Lorrain's *Apollo Keeping Sheep*.[73] But he has been the happy captive of architecture up to this point, ever since the approach framed the great house like a Claudean canvas. He finally leaves the house for the fields.

Young had no doubt that the splendor of Holkham (and thus the power that could be expressed through landscape) derived from the muck in its fields. He claimed that before the events of the eighteenth century, Holkham had been nothing more than "a wild sheepwalk before the spirit of improvement seized the inhabitants; and this glorious spirit has wrought amazing effects." Instead of being full of wilds and wastes, "the country is all cut into inclosures, cultivated in a most husband-like manner, richly manured, well peopled, and yielding an hundred times the produce that it did in its former state."[74] At Holkham they laid twelve loads of dung on every acre, marled the ground, broke it up for wheat the first year, and from there followed the four-year shift of that country—turnips, barley, clover, and grass. Young described the root crops for the animals, the pens where cattle made manure, the many duties of the lord's tenants, the yields and the incomes that brought the rents that made for lavish houses and great estates. Nothing but a sheep walk, avowed Young, transformed by the master of the house and others like him all over England: "How, in the name of common sense, were such improvements to be wrought by little or even moderate farmers! Can such inclose wastes at a vast expence—cover them with an hundred loads an acre of marle—or six or eight hundred bushels of lime—keep sufficient flocks of sheep for folding. . . . No. It is to GREAT FARMERS you owe these. Without GREAT FARMS you never would have seen these improvements."[75] A gross overstatement at best. What Young did not seem to know was that the so-called great farmers only extended innovations in use for a century before the Earl of Leicester tried them on the sandy soils of Norfolk.[76]

Yet the entire effect of these practices, duplicated on estates throughout the English countryside, had no parallel in the eighteenth century. What jumps from the pages of British agronomy are the startling changes to the environment that intensive expansion brought. Throughout the century lasting from 1750 to 1850, landlords and tenants turned up grasslands, replaced old arable with pasture on a fantastic scale, grazed sheep on three-quarters of the surface of the island, reclaimed heaths, planted moorlands, planted hedges and extended enclosures, drained marshes, cut woodlands, and brought at least two million acres of waste under the plow.[77] This list includes some of the defining practices of the Agricultural Revolution. John Sinclair advanced the idea that landlords could radically alter the ecology of their farms, and even the overall climate, through bold measures over patterns of land use. "The climate of an extensive region," Sinclair wrote in *The Code of Agriculture*, "is improved, by cutting down large forests, by draining great lakes, or extensive marshes; and above all, by judicious cultivation."[78] English agriculture drove these changes because it drew razor-sharp lines between waste and enclosure, between this side and that side of the hedge. The entire system just described allowed farmers to concentrate ecological functions in an enclosed space where a cycle could be established and where nothing with the potential for havoc would be allowed. Fire and fallow threatened the tightened control landlords attempted in agricultural spaces by ignoring new boundaries—fire by erasing them and fallow by inviting any seed on the wind to cross them and take root. Chaos would be leashed, and the nutrient cycle represented by manure would create absolute permanence, perpetual cultivation on the same ground without fallow and without end. The system offered ecological stability in a capitalist mode of production, and it would not be disturbed before the onset of industrialism in the British countryside.[79] Americans did not draw lines in the landscape quite as sharply, but the same ideas appealed to farmers in the old states confronting their own unruly commons in the West, looking for ways to capture the same stability and abundance.

Americans who reached out to British husbandry saw it as the buttress of their wealth and the perfection of their settlement. Jesse Buel never tired of advocating the manure religion: "Farmers should hence regard manure as part of their capital—as money—which requires but to be properly employed, to return them compound interest. They should husband it as they would their cents, or shillings, which they mean to increase to dollars." Writers of lower distinction than Buel said the same thing. According to "A New Theory of Agriculture" (1821), one of hundreds of works on the subject, "Every operation of husbandry, every preparation of land, is calculated to render manure efficacious in its application. . . . In fine, it is that part both of the theory and practice of agriculture upon which every other may be said to depend." For John Lorain of Pennsylvania, convertible husbandry carried a kind of charisma that was the extension of the restorative hand of the farmer himself: "The sight and smell of a fermenting dunghill . . . quickly demonstrate the course that should be taken with this invaluable article, for when fermentation takes place beneath the soil, the fructifying and exciting properties of the manure are diffused through the whole mass, and nothing is lost which could have been possibly saved."[80]

Yet the religion also had its doubters, people who insisted that it did not fit the social or environmental conditions of an expansive country. John Hare Powel, a Pennsylvania cattle breeder and a dreaded gadfly among his peers, liked to sling rhetorical dung in the faces of Anglophile farmers. "On the Evils of Soiling in a Country where Land is Cheap and Labour is Dear," he expounded, "I am confirmed that in *this climate*, soiling can seldom be profitable."[81] By "soiling" he simply meant the meeting point between cattle and their well-selected fodder—the wintertime pen where the animals turned out dung for their keepers. Powel argued not against the system itself but that Americans either got it wrong or found it too expensive, and he insisted that cattle failed to thrive in confinement: "The bloated, sleek, and pampered calves, which we have seen taken from soft beds, and dark stables, to be dressed in ribbons or exhibited at shows, are well fitted to deceive dillettanti [*sic*] farmers, or to decorate the butcher's stall," but they would never be hearty.

More serious than feeble cows was the money out-of-pocket that farmers had to spend if they wanted to impersonate English landlords. It could be done, Powel believed, but only specialized cow keepers and small freeholders "may possibly, if they have the vigilance of New England, succeed." He doubted that south of the Hudson River the balance between land and labor permitted convertible husbandry in any form, for "whilst the richest meadow pastures can be had at $8 per acre a year, and labour can be procured but at the high prices . . . no man can profit on a large scale by the system, which in England, has been made successful, by the *cheapness* of labour, and the *high price* of land." As for the apparent simplicity of the methods, Powel balked at that too. One reason why many rejected the high husbandry was that it required individual experiment to figure out what worked in a given location. "Hence the great prejudice in the minds of the lower classes of society, against the most improved systems of cultivation," wrote Richard Buckner in the *American Farmer*. Some raised their voices at the idea that Americans might suffer high costs while bowing over the Atlantic. "Your farming is now all done, and I trust well done," said one manual. "No man has thrown away a dollar unnecessarily upon new and visionary schemes, by making experiments upon *English* farming in *our* country."[82]

British and American farmers may have agreed on the means, but they saw the results very differently. Americans tended not to quibble about or even compare yields, as though they were beside the point. If the new system failed to increase productivity by enough to justify Arthur Young's more sensational claims, that hardly mattered to Americans, for whom any increase was notable.[83] British landlords considered yields of twenty-two, twenty-three, and even twenty-six bushels of wheat per acre good but unspectacular, while Americans cheered when that much wheat headed up. One report on domestic grain production estimated the average product of the United States at fifteen or sixteen bushels to the acre, while Germany, England, and France averaged from twenty-four to twenty-six.[84] More often than not, big harvests in the United States matched normal levels in England. So although Young may have overstated the power of improvement to generate wealth for the aristocracy, his work carried a different meaning in North America. Recall that farmers usually emigrated not because they had demolished their topsoil but because their land no longer brought harvests and profits sufficient to allow them a modicum of comfort. Improvement offered not riches but stability. For those who stayed east, farming had to be a fair investment, offering at least the possibility of increasing returns, and had to be conducted on land that would never need to be abandoned or sold. That is all the dunghill doctrines promised.

In that simple understanding lay a conception of the old states not as gullied waste but as renewable resource, as the spring of American expansion rather than a casualty of it. Worry over oldfield and depopulation obscured another emotion—an optimism about the capacity of new practices to make the domesticated earth even more plenteous than first nature. Said one author writing for New York schoolchildren: "Instead of farms growing poorer as they grow older, as has generally been the case in this country, it is now discovered that they may be made to grow richer."[85] Samuel H. Black of Delaware insisted that most farmers sleepwalked for decades "over a mine of wealth, and yet die leaving posterity heirs only to their wretched poverty." Though many sold their land at less than $20 per acre, "or are suffering the sheriff to do it for them," Black concluded that "every acre of it which is arable, whether it be now rich, or poor, is intrinsically worth *five hundred dollars*." Improvers believed that a generous influence over nature lay in their hands, with which they could create an affluence of fertility leading to the financial kind. Or as John Taylor wrote in a moment of optimism, "The farm well managed according to this system, will, in twenty years at least, return back to its original fertility."[86] This is the restored faith of a planting people.

Another point of view gaining ground in the 1830s and 1840s combined an optimism about nature with an overarching pessimism about culture. Nature romantics depicted wilderness as morally superior for being desolate and regarded settled places as morally ambiguous or, worse, fatally corrupt.[87] This philosophy contained a critique of rationalism that would have enormous influence throughout the next century, yet it had little to do with actual environments and little to say to those working day to day to manage the disturbances of agriculture. Henry David Thoreau thought all the "many celebrated works on husbandry, Arthur Young among the rest," totally missed the point:

> If one would live simply and eat only the crop which he raised, and raise no more than he ate, and not exchange it for an insufficient quantity of more luxurious and expensive things, he would need to cultivate only a few rods of ground, and . . . it would be cheaper to spade up that than to use oxen to plough it, and to select a fresh spot from time to time than to manure the old, and he could do all his necessary farm work as it were with his left hand at odd hours in the summer.[88]

Even the richest land exacted the labor of a lifetime in exchange for a life of middling comfort, and all husbanding people knew it. Yet Thoreau dismissed this deliberate economy and asserted that any farmer could shift around indefinitely while tilling a small parcel, and do all this with little effort, "with his left hand." Easy for him to say. Thoreau engaged in a thought experiment, camped out in his backyard. He testified brilliantly that people could strip away the suffocating matter that intrudes on a philosophical life, yet he assumed too much about the farmers he referred to as his neighbors. Whoever they were, they lived more squarely than he did at the intersection between economy and ecology.

True, some people had begun to question the meaning of progress. Henry Thoreau's asceticism and Thomas Cole's dismal prophecy in *The Course of Empire* (1833–1836), an epic painting in five panels depicting civilization's fatal violation of moral and natural parameters and its inevitable destruction, resonated among those who believed that wildness formed an invariable standard. Romantics invented "Nature" as a refuge from "Civilization." Farmers, on the other hand, had no time and no mind for the stark dualities of these intellectuals. Improvers may have composed a minority of all farmers, but they certainly represented the majority culture, with their cheerful conviction in the redemptive power of human hands on the world. Convertible husbandry rejected the idealization of nature by arguing that the full richness of land could only be realized by aggressive human manipulation. The earth did not wilt when touched by people; people wilted from their duty to take good care. These differences, however, obscure what Thoreau and his more conscientious neighbors held in common. Away from the extremes, romantics and improvers grappled with the same problem: how to rescue stability in nature during a time of accelerating change. Many romantic thinkers embraced the middle landscape as an alternative to wilderness, and Thoreau himself felt more comfortable in his bean patch than he did climbing Mount Ktaadn.[89] Paintings by George Inness and Frederick Church, especially the graceful *Haying near New Haven* (1849), confirm the countryside as civilization's steady state. Improvers helped create the very landscape of beauty, compromise, and stability that romantics idealized, seeing it as a realm in which a commercial society might foster both garden and forest.

From this point on, the dunghill doctrines lose whatever unity of purpose they might appear to have inspired among farmers and planters in the old states. The nutrient cycles and grassy courses of convertible husbandry never had a single meaning and never represented a single politics; and though James Madison chose not to consider, or even acknowledge, regional aspirations in the restoration of land, others thought of nothing else. No improver in Britain or the United States ever doubted that the reform of agriculture had political implications.

Two

That the strongest chord which vibrates on the heart of man cannot tie our people to the natal spot, that they view it with horror, and flee from it to new climes with joy, determine our agricultural progress to be a progress of emigration, and not of improvement; and lead to an ultimate recoil from this exhausted resource, to an exhausted country.

—John Taylor (1819)

In many of the settlements around us, the natural fertility of the soil has been exhausted. Comparatively little new grounds remain to be brought under cultivation. To refertilize the old must be our resource; and the necessity is every day becoming more imperative.

—The Pennsylvania Agricultural Society (1823)

Island States

People moved, and free people moved often. Thomas Hart Benton of Missouri, first among boosting western senators, orated relentlessly for a reduction in the price of public lands to bring about the rapid transfer of population from the states to the territories. He might have put it this way: the United States, having purchased Louisiana, had no right to own it. He sermonized on the western lands in 1828, saying, "We must rouse them from their dormant state. We must infuse new life and animation into the sales. We must give them an accelerated movement in the path of their original destination." Open territory, once taken from Indians, had no other purpose than to be dispensed of by government as a stage in the formation of new states, "giving them the use of all the soil within their limits, for settlement and taxation—multiplying the number of their freeholders—raising many indigent families from poverty and wretchedness to comfort and independence—and converting some millions of acres of refuse and idle land into a fund for the promotion of the great cause of education and internal improvement."[1] Benton never ran out of words for the monumental importance of the public domain, and he even cast the material progress and political survival of the United States as propositions that depended on a liberal policy.

Improvers never looked at the West this way. They emphasized community diaspora, not community formation, land abandoned, not land found. But what was it that led them to meditate so darkly on the common desire to find a better life by moving on? Emigrants crossing from South Carolina into Georgia scarcely noticed the border, but to the people who remained it marked a kind of coastline.

Shifting populations reapportioned political capital because, according to the Constitution, fewer people meant smaller delegations to the House of Representatives. This came at a critical moment—when the northern states moved against the spread of slavery into Missouri, when rising numbers of manufacturers had begun to distinguish the economic interests of New England and Pennsylvania from those of Virginia and South Carolina. Southerners who called themselves War Hawks in 1812 and nationalists in 1815 had already made motions toward sectionalism by 1820, so the balance of population in these semi-sovereign entities became a factor in the balance of power within the Union. It is easy to forget that the Revolution and the Civil War bounded the era of preeminent states. The slavery issue rendered them like great European powers and Congress like a stormy summit to decide the fate of the continent. The great political personalities of the times seemed to grow from them like limbs: Clinton of New York, Calhoun of South Carolina, Webster of Massachusetts, Jackson of Tennessee, Madison of Virginia, Clay of Kentucky. The *States United* better describes national politics before the victory of the Union nullified the Tenth Amendment, in which all powers not "delegated to the United States" belonged to the states or to the people.

Here was the fear of limitation in a time of unimaginable land abundance: states were islands in a political sense. Keeping farmers in place or losing them over an invisible line moved scales miles away in the national legislature, so the people most interested in improvement tended to be those with the deepest roots in the soil of their states.[2] Northern improvers hoped that walling in the old thirteen would keep their states in office, so to speak, as the elder statesmen of the Republic. Not only did movement west slow the growth of the ancient commonwealths; it seemed to challenge an older order in which East dominated West. The sensibilities of many reformers strongly resembled Federalist attitudes toward an earlier frontier, especially the desire to assimilate the territories at the same level of "civilization" characteristic of the East. As they might have put it, no society founded on a rising land bubble and a reckless brand of western democracy would ever attain the moral authority of a well-constituted state.[3]

George Perkins Marsh embodied this habit of thought. Marsh was born in Woodstock, Vermont, in 1801 and came of age during a time of rapid environmental change in New England. In 1864 he published the foundational work of American conservation, *Man and Nature*, in which he connected erosion and deforestation to the waste of land. Years earlier, Marsh called emigration a misstep as much for environmental as for political reasons. His own inspection of the West in 1837 ended bluntly at the Falls of St. Anthony, present site of St. Paul, Minnesota. He mistakenly called most of what he had seen along the way sterile and uninhabitable, and he never went back. An early biographer supposed that the privations of removal would have been "fatal to his best intellectual growth." Marsh had entrenched interests in Burlington and loftier stuff in mind than gross acreage, like the greater imprint he might make by remaining at home, for Marsh "regarded New England as the mother who was chiefly to form the character of the rising States of the West."[4] Nations developed not by adding territories and cultures that could not be assimilated to the whole but rather by cultivating internal qualities. Marsh measured change in long-settled places and believed that civilizations flourish to the extent that they abide. Not all rural reformers resisted expansion with Marsh's immovable certainty, but most who considered the problem believed that expansion encouraged the illusion of abundance and that led to waste. Like all consumption, expansion needed brakes, self-regulation. This is the crucial link between improvement and conservation.

Southerners thought differently about the meaning of land and restoration, and no one advanced their view with the same strange vigor as John Taylor—planter of Caroline County, friend of Jefferson and Washington, and perhaps the most ardent voice of the old republicanism in print. Taylor published sixty-four essays beginning in 1810, collecting them under the title *Arator* in 1813. *Arator* elevated him to great status among American farmers, for whom he wrote a hard pastoral, insisting on agriculture as moral instruction, not Arcadian escape.[5] Think of Taylor's epistle as an anti–*Federalist Paper* to counter James Madison's constitutional rendering of nature. In order to survive in a political world made dangerous by a centralizing government and an industrializing economy, the plantation needed an agronomy of retrenchment and uncompromising purpose.

The immediate political context of *Arator* consisted of the raising of "bounties" or tariffs by Congress in support of domestic manufacturing. Under tariffs, finished products from abroad paid duties before entering the domestic market, thus raising their prices in order to discourage Americans from buying them. Meanwhile, on the other side of the transaction, farmers compelled to purchase foreign goods accused the government of gauging them so that manufacturers might enjoy an advantageous home market. Taylor called this "bounty" for factories an unconstitutional tax, a fatal corruption of government, and a "bribe" offered to farmers in the dubious form of lower domestic prices intended to pacify them as they became "dependents on a master capitalist for daily bread."[6] No good feeling for him, Taylor denounced the American System (the first national program to finance and construct roads, canals, and other internal improvements) as though it were a boot heel against his neck. With this, Taylor's argument comes to a point: by taxing planters, tariffs made further investment in land more burdensome. Rather than invest what money they had in livestock or feed, planters wore out their soils and packed for the state line, leaving behind a waste and a brooding instability that were Taylor's spurs to write *Arator*. This is the entrance to Taylor's political agronomy. Only a political solution would save the Constitution and prevent government from killing what gave it birth, but only an agricultural solution would save the South.

Arator derived the elements of a more perfect plantation—a fortress against the chaos of the times—and convertible husbandry formed its immovable foundation. Taylor referred to the improved agronomy as "inclosing" for the way it separated cattle and grass in order to foster the growth of both. The plantation needed to thrive in its autarky because it anchored the "nation"—an organic, almost mystical association of white freeholders that existed apart from any government. The nation occupied no particular state or section. It certainly owed nothing to a messy constitution but to birth, place, blood, and soil. The nation existed by nature, and agriculture served it fiercely: "The nation never dies; it is the yoke fellow of the earth; these associates must thrive or starve together; if the nation pursues a system of lessening the food of the earth, the earth in justice or revenge will starve the nation. The inclosing system provides the most food for the earth, and of course enables the earth to supply most food to man." The nation's only weakness came from the disheartening desire of its people to leave it. "Agriculture can only lose its happiness by the folly or fraud of statesmen, or by its own ignorance," Taylor said.[7] The second was much more difficult for him to face, and it proved the central problem of rural reform in the South.

Marsh and Taylor belonged to different places, and they drew the connections between nature and society in completely different ways, yet both dreamed of an American *Landschaft*. The word can be translated as province, district, region, or countryside; it carries the suggestion of sufficiency and constancy. *Landschaft* also implies a continuum in the landscape, with a core of nestled buildings and gardens surrounded by a border of fields, then pasture, then wilderness. *Landschaft* tells an agrarian fable in which communities grow ancient without deterioration, in which people never ask more from their immediate environment than it can give. The idyll never fit American conditions, yet it more or less captures the way that improvers thought about their farms, plantations, and communities. The fragility inherent in *Landschaft* is the fragility certain farmers sensed when they confronted soil depletion and widespread emigration.[8] And because *Landschaft* told a story of home, it was an idea worth fighting for. When they spoke about their fears of social decline and disintegration, improvers made political statements.

This chapter depicts some of the political implications of the new husbandry in two states: Pennsylvania (where it found its greatest New World expression) and South Carolina (where it ran up against plantation slavery and a poor environment). The story is all about boundaries—marking off settled and unsettled, East and West, affluent and poor. Pennsylvania farmers and South Carolina cotton planters extended the idea of "inclosing" from the edges of pasture to the boundaries of their states in an attempt to create order at home and quell the popular desire for unbroken land. They did not always speak of state and politics, but in every instance improvers used agriculture to shape a place of strength and continuity. But how important was that place, anyway? Was agriculture enough to protect the influence of the old states? In an expanding Union was it possible to base a politics on such old ground? Or was improvement just a way of sharing the anxiety of change wrapped up in a hopeful rhetoric about the homeplace? Emigrants haunted these questions because, like soil, they moved about and carried the future fertility of the nation with them.

ANOTHER WORLD

The emigrant highway began in southeastern Pennsylvania. The world of woodcraft, axmanship, log cabins, and subsistence hunting first formed in the lower Delaware River valley, just below Philadelphia, in the seventeenth century, and from there its core domain extended through the Cumberland Gap to Missouri and as far as the German Hill Country of Texas. The occupation of the North American forest by a definitive backwoods culture exploded from Montgomery and Berks Counties after an infusion of Scotch-Irish and German immigration in the 1720s.[9] Tens of thousands of people continued to cross the commonwealth a century later, many of them over a road that extended from the foothills of the Allegheny through Bedford to Pittsburgh. Others came by way of the Erie Canal to Buffalo, and then down the lake shore. The National Road began in the town of Cumberland, on the southern border, and cut through the entire northwest, by way of Indianapolis, almost to St. Louis.[10] Unlike the New England states, though somewhat like New York, Pennsylvania had a long familiarity with the West for the simple reason that it had one of its own behind the mountains. The borders between nations and states often follow the incline of ranges, but Pennsylvania somehow managed to hang together even with the Allegheny, the northern portion of the Appalachian Mountains, slicing off the southeast from the rest of the state.

Philadelphia, located on the Schuylkill above Delaware Bay, in easy communication with New York and Baltimore, was the capital of the eastern state of Pennsylvania. Farms in the bordering counties of Bucks, Montgomery, Chester, and Lancaster sat on northern Piedmont soils in a region where farmers of German and English descent worked some of the finest farmland in North America, where rivers and streams flowed rapidly on their way to the coastal plain, forming ideal locations for water-powered mills. Pittsburgh was the capital of the western state of Pennsylvania, in the watershed of the Ohio River, closer in miles and temperament to Cleveland than to Harrisburg. It is the only city on the Appalachian Plateau, sitting amid a confusion of river valleys and low hills unfit for prosperous agriculture.[11] The affluence, urbanity, and industry of the state's southeastern portion made its inhabitants different in outlook from the more recent settlers of the mountains and the West. The train of movers caught the attention of farmers and others fearful of great shifts in population and civilization during the depression of the 1820s. Their anxiety over emigration came out in frustrated addresses and treatises intended to hold the movement or, if that failed, to distinguish East from West, affluent from poor, old state from new. The shore of Pennsylvania's island state adhered not really to its political borders at all but to the shifting line between the settled places and the "western wilderness," the line between two agricultures and two rural societies.

Easterners watched in amazement as people traded a modicum of comfort and familiarity for struggle. Evans Estwick set out from New Hampshire to see what all the excitement was about. Little satisfied him: "Some have been so imprudent as to abandon the home of their infancy, where the comforts of life could have been obtained by a good degree of industry. What were the consequences? perhaps wealth;—but it was unnecessary;—perhaps poverty, disease and premature death. Some too, even in advanced life, and after spending their prime in clearing a tract of land, so as to render it fertile and easy of cultivation, have sacrificed a comfortable and pleasant old age for new perils and labours in the western wilds." People who could possibly live well at home, he said, "will act wisely in remaining where they are."[12] According to a Vermont pastor interested in the recent removals: "For several years, the moving mania was so great, that it threatened almost to depopulate whole sections of the country. . . . Although the soil, in almost every square mile in the county, is like a garden, numbers have seemed to think that it was nearly exhausted, and that it is the part of wisdom to seek a 'newer country and better soil.' But the error is found in the judgment of the occupant, and not in the soil on which he treads." With no way to take a census of domestic movers and newly arrived immigrants, easterners sometimes became frantic. Indiana and Illinois, where northern roads pointed, grew from a combined population of 202,389 in 1820 to 500,476 in 1830, or by 68 percent.[13] And though only one in ten of the people headed for the prairies came from the North (most came from Kentucky, Tennessee, and upper Virginia), denizens of the eastern countryside noticed that European immigrants landed at Philadelphia, passed through the better-cultivated counties where land cost too much, found their way to Erie and Pittsburgh, and kept right on going.[14]

Startled by the threat of changing fortunes, the Philadelphia Society for Promoting Agriculture addressed the citizens of Pennsylvania in 1818. The erosion of agriculture was creeping into the erosion of society, they feared, as farmers fled debt and bad lands:

> The torrent of emigration is increasing daily, and its effects on the countries in which it takes its rise, are already disastrous and fatal. In some parts of what are called the old states, whole towns or townships, are almost without an inhabitant, and former proprietors having abandoned their farms, attracted by the glare of the new countries; and while the present bad systems of Agriculture, which were practiced in these districts prevail, while the productions of the soil, from improper modes of tillage continue annually to diminish, these emigrations must and will increase.

Nearby, in New Castle, Delaware, Samuel H. Black also spoke worriedly against the trend and urged his fellow farmers to discover "the intrinsic value of arable land":

> Already we see crowds of emigrants pressing from the worn out soil of the Atlantic, to the fresh lands of the interior, and western States. The result will be, that these States must be abandoned by the most honest and worthy of their inhabitants, or a total, a radical change must be effected in the management of our land. . . . It is as imperiously the duty, as it is obviously the interest of every man in these States, who possesses anything which attaches him to his natal soil, to join heartily in the effort of retrieving the credit, the prosperity, and the wealth of his neighbourhood.[15]

The statements read as testimony to the speakers' geographical and social distance from the lives and needs of practical farmers. The Philadelphia Society for Promoting Agriculture attracted merchants and attorneys with an intellectual or political commitment to rural life. Tension within the organization almost always came from the rift between members who resided in the city and those who operated farms nearby. The sense that theory existed for its own sake and not to change the way farmers attended to soils finally tore the PSPA apart after a clique broke away, intending to establish a new organization dedicated to "practical husbandmen," not the powdered type.[16] Yet when it came to the log-and-hatchet world of emigrants, few farmers of the southeastern counties knew very much unless they had seen it themselves. Few had. In the southern states planters of all descriptions contemplated the West, and even some of the wealthiest owners of slaves and cotton packed up for Alabama, but in the North the border between those who stayed and those who removed more closely followed the contours of class.[17] A stake in state and neighborhood, a sentiment for land well established and passed down, a belief that cultivation stood as "the grand pillar upon which rests the whole civilized world," comparable to every other pursuit like "what the sun is to the planetary system, the light, the heat, and the soul of all"—these positions defined a certain group of people for whom removal represented failure.

The subject is vexed by facts on the ground. A people watching a torrent of emigration leave them poor and isolated might look to their bad cultivation and repent, but Pennsylvania gave every sign of sustained and vigorous growth. In 1810 it counted a population of 810,019. Ten years later 727,977 people lived in the eastern portion alone, with a total of 1,549,458 residents, representing a 90 percent rate of growth—the most populous commonwealth. Pennsylvania emerged from the War of 1812 with one of the strongest economies in the United States, a hive of manufacturing large and small, with perhaps the fattest countryside in the Union. People went great distances *just to look at it*. A southern journalist from one of those states where farmland had been "turned into desert places by miserable cultivation" thought he had landed "in another world." He had landed in Bucks County, where lands sold for as high as $50 and $75 per acre, even during the depression. He found farms of 180 to 200 acres (only a tenth of them in wood) where the yearly production typically included 1,000 bushels of corn, 350 bushels of wheat, 1,200 of oats, 100 of rye, 300 of potatoes, fifteen steers of six hundred pounds each, and three thousand pounds of pork raised for sale. Labor employed in orchards and cider houses earned $300 a year.[18] The look of the people bespoke the look of the land: red faces, hard hands, people said to be broad and hearty, with milk and butter in their blood.

Not good practice alone but a natural endowment made southeastern Pennsylvania a remarkable countryside. During a year of miserable yields a traveler from Delaware passed from Maryland into Pennsylvania and watched the wheat harvest increase from one to thirty bushels per acre inside the state line. The visitor called on a Lancaster farmer who vaunted that at least part of his property had never seen a load of manure—not in thirty-six years—and claimed that it might still give forty bushels.[19] The author of a book boosting the scenery and internal improvements of the state could have written about agriculture anywhere but chose Lancaster: "It is, without doubt, the garden of this glorious Union, and there are few spots in this wide, wide world, which could present a nobler scene to the eye than is here afforded. . . . The entire region [of the Cumberland Valley] presents one continuous and almost unvaried scene of agricultural prosperity."[20] Finally, the English journalist Henry Bradshaw Fearon, who hardly ever had a kind word for farmers on these shores, paused in the same Dutch country and delivered the ultimate compliment: the "superior cultivation of the 'Great Valley,' place[s] it decidedly in advance of the neighbouring lands, and put[s] it fairly in competition with Old England."[21] The endowment of the counties proximate to Philadelphia and the gross disadvantages of other regions will come up again, but a general problem should be noted now: practice cannot be divorced from the absolute qualities of land that circumscribed every farm. Any smugness on the part of Lancaster farmers should be discounted at once. They did not always trace a finer till than others on less favorable land; but what improving methods they did attempt yielded better and more to the hand.

A single structure symbolized all the prosperity on the Atlantic side of the Allegheny: the Pennsylvania barn. After the steamboat and the Erie Canal, this building, in its size and capacity, made visitors marvel at American ingenuity and abundance. During his year at large in the United States, William Cobbett, a British political observer, wrote about the rural districts not far from Philadelphia, calling them "a fine part of America," with "Big barns. . . . Barns of stone, a hundred feet long and forty wide, with two floors, and raised roads to go into them, so that the wagons go into the first floor up-stairs. Below are stables, stalls, pens, and all sorts of conveniences. Up-stairs are rooms for threshed corn and grain; for tackle, for meal, for all sorts of things. In the front (South) of the barn is the cattle yard. These are very fine buildings. And, then, all about them looks so comfortable, and gives such manifest proofs of ease, plenty, and happiness!" The barn first appeared in eastern Switzerland and western Austria, definable by its cantilevered overhang known as a forebay. The stone masonry girding its giant foundation could be sighted half a mile off. Nineteen percent of the houses and 4 percent of the barns in New Castle County, Delaware, were made of stone in 1798. By 1828 the portion had increased to 45 percent of the houses and 30 percent of the barns.[22] The barn functioned as a macrocosm of the cow, a human superstructure over bovine digestion where fodder crops entered one end and milk and manure came out the other. Where other farmers called manure a nuisance, the Pennsylvania Germans reserved it like gold.[23]

James Pemberton Morris built a barn on his Bucks County farm in 1823–1824. With his membership in the county society and his neat daily journal, Morris was a practical improver, one who attempted to join bookish ideas to the necessities of dirt and rain. Without question, his farm represents the apex of northern agriculture in its time.[24] The journal maps the diversity of farmscape and the complexity of good management that reformers urged on all plow-people. It is worth simply listing the products that could be found on Morris's place, more or less in the order that he mentions them: turnips, merino sheep, dairy cattle, clover seed and hay, apple trees for fruit and cider, oats in fields and in the young orchard, potatoes, wheat, drumhead cabbage and early York cabbage, radishes, English turnips, carrots, parsnips, beets, peach trees, buckwheat, rye, timothy. He had the essential improvement crops, the roots and grasses that he used to create highly productive meadow, and he distinguished between dairy and other cattle, a sign of his attention to breeding. Notice that, like Morris himself, I want to mention productions along with places on the farm. Morris created interlocking relationships—a rural ecology—from what might look like scattered fields, woods, and yards.

At first plowing in March 1823, Morris used three horses to cut the clover sod, where he planted corn, and did the same in another field where he put in wheat. In the meantime, his hired hand threshed clover seed, probably from hay harvested the previous summer and stored. Morris sowed clover with corn and wheat, fertilizing his land and planting it at the same time. There was still snow on the ground, so clover sod represented the earliest possible planting for the season. Any rocks pulled up went to the foundation of his new barn. He planted oats in the peach orchard and buckwheat in the spring meadow so his cattle could feed from the straw and stubble after harvest and manure the ground where they ate. Morris kept a "pattern field," where he may have conducted experiments, though he seems to have used it for whatever need presented. "Six hands mowing are making hay" on the "hill field" in June. Morris used ground too steep or rocky for regular cropping, requiring little or no care to maintain, as a way of enlarging his fund of fodder crops in addition to his seeded meadow, as though permanence in the agricultural landscapes required a remainder of casual flexibility.[25] Every farm needs a less cultivated peripheral space where more is taken than given, perhaps a place where fire is permitted, generally where the regular order does not hold. Here sheep could fend for themselves or the hay could be cut.

Stand back from the detail, and the essence of the northern improved farm comes clear. Morris lived at the epicenter of animal husbandry in the United States, where domestic quadrupeds represented the predominant source of land use. There is no way to determine the size or number of his fields, but Henry Bradshaw Fearon estimated the acres typically enclosed in the same neighborhood. "In a farm of 200 acres," wrote Fearon, "the proportion may be estimated at 90 acres of ploughing, 50 of meadow, 10 of orchard, and 50 of wood land. . . . A farm of the above description is worth, if within five miles of the capital, 20,000 dollars; at from 20 to 40 miles' distance, 10,000 dollars."[26] Three-quarters of the non-wooded area in meadow, or one-quarter of the total farm, is impressive, but it only sketches the picture. Neat categories of land use obscure the true influence of animals. Every clover and hay crop, every patch of grass and corn stubble, any root crops not in the garden, the debris from threshing and apple picking, along with mast from the woodlot—all these fed animals. In other words, virtually every place on the farm, at one time or another, might host a ruminating cow or ewe. On top of that there was the barn, a building to provide inside space for animals and their food.[27]

Such a gracious dedication of resources did not go unnoticed. John Hare Powel, the cattle breeder, deduced from his own survey that no more than one-third of the land in the eastern, middle, and western states was plowed annually, and "of its produce, except small quantities of hemp and flax, the farinaceous parts of wheat, buckwheat, and a portion of rye and Indian corn for whiskey and bread, nearly the whole is employed for the nourishment of Neat Cattle, Horses, Sheep, and Swine." Powel quoted another English agricultural theorist, John Sinclair, to support his implicit claim that a predominance of meadow typified a long-settled country: "Probably not more than two-fifths of even the arable land, or ten acres in a hundred of the whole surface, produce crops immediately applicable to the food of man. The remaining ninety acres, after a small deduction for fresh water lakes, are appropriated to the breeding, rearing, and fattening of live stock." The proportion seemed to extend to New England, where an observer of land use reported that "nine tenths, some say, nineteen twentieths of the land in this County, setting aside wood land, are used for feeding live stock."[28] Forty percent of the text of *The American Farmer's Instructor* by Francis Wiggins (published in Philadelphia in 1840), one of the most lucid works of its kind, covers animals, and Wiggins concludes: "The feeding and fattening of cattle, whether for labour or for sale, is the most important of the whole economy of the grass farm."[29]

The 1840 census records the footprint that cattle made on the northern countryside. Even allowing for the inevitable mistakes and misrepresentations of the census, the differences between North and South are astonishing. The region with the smallest total area, the Northeast sustained more cattle than the Southeast, Southwest, or Northwest by 10 percent, attaining a density double that of the Southeast (thirty animals per square mile versus sixteen). New York and Pennsylvania alone accounted for as many head as the five oldest southern states. Cattle form only one side of the equation, of course. With an average production of close to fifty tons of hay per square mile, representing 77 percent of the nation's total, the Northeast was a wonderland of feed, where broad lowing creatures masticated over a great deal of the landscape.[30] Does the presence of hay and cattle together in the same places prove that farmers practiced convertible husbandry or that they made manure? Hay went to dairy cattle to make butter and cheese, the most highly valued commodities in the rural North, and to horses and oxen to provide them with work energy. Farmers might not have considered manure of *primary* importance, but keep in mind that they never set out to do just one thing. Every part of the farm implied every other, so the idea of a by-product made little sense to them. Every three head of cattle in the Northeast had five tons of hay to eat, all in the same square mile. Farmers who went to the trouble of cutting the meadow had an incentive to make the most of the labor, time, and energy they expended. The way to do that was to pen up their cattle and reap a second harvest.

An elegant document helps to make this argument. *The Cornell Farm* (1848), a folk painting by Edward Hicks, records the buildings and stock belonging to James C. Cornell of Bucks County. We stand at the foot of a large yard framed between a substantial house and an even larger red-and-white barn. Orchards stand in the distance. Ten people look and gesture, including a hired hand pushing a plow behind a horse and a man in white, perhaps the proprietor. But never mind all that—the purpose lies in the foreground. More than forty animals strut and pose in individual portraiture, winners at the annual cattle show sponsored by the county agricultural society. Cattle in four or five breeds, built like strongboxes, udders full; horses of the same number, muscular and vigorous in gray and white; black pigs; and sheep like walking cotton balls. The entire farm exists in their light.

These are the elements of a thriving rural landscape and also of a busy commercial society, where farmers managed not only multiple commodities but remarkable cash flows. In May 1823 alone Morris employed no fewer than three carpenters, four masons, two tenders, and four hands. The men who hauled his stones probably owned no land of their own, and without them he could not have built a barn. They also mowed his meadow, went back and forth to the sawmill, and planted his corn next to the woodlot—allowing Morris the freedom to attend market, address his fellow husbandmen, and conduct a business that brought him into contact with banks and agents. The paper he handled distinguished Morris as among the most capitalist of farmers in the United States. All at once within the same few days, he paid out wages to his workforce, lent $150 to a relative, rented out property for a term of seven years at $200 a year, and paid a debt of $700 to another relative through the local bank. Intensive cultivation and market connections mutually reinforced each other; the one made the other necessary. In October, Morris sold one hundred bushels of apples as an "adventure" to Washington, no great distance for a bale of cotton but a trip that pushed the limits of local-supply agriculture and marked the depth of Morris's ambition.

Yet the world of the high-end Pennsylvania farmer was by turns commercial and communitarian. The cattle show where the farmer Cornell won premiums for his stock and the people milling around in the frame suggest a community of interests that circumscribed his land and property. County agricultural societies published addresses and sponsored experiments that spoke to individuals and their private pursuit of wealth; as leagues made up of neighbors, however, they had an unmistakable mission to encourage the public behavior that would most likely guarantee collective security.[31] Farmers spoke explicitly of the common good emanating from their practice. No husbandman reeking of the yard late in the day looked "so sleek as dancing masters," but, as a correspondent to the *Carlisle Republican* enjoined his readers, "you are doing more good—you are doing something to increase the common stock—the means of subsistence—the real wealth of the nation." Capitalism had not yet supplanted traditional ties in the 1820s, and James Pemberton Morris is again good evidence. He might have *built* a barn with wages, but he needed more people than he could possibly pay to *raise* it: "75 neighbors assembled for the purpose of raising barn roof, which we accomplished easily by 5 o'clock and nobody hurt." And though Morris and his neighbors bought and sold much too much to be mistaken for true yeoman farmers, a good portion of the dollars passing through his hands in May 1823 ended up in households linked to him by kin.[32]

Economy and society were not distinct spheres, but agronomy gave them unity and stability. Swirling around the theory and practice of sophisticated animal husbandry came an ethic of consumption never before seen in the United States—a mixture of upper-class aspiration, eastern chauvinism, and love of place.

How do we connect the farm in the shadow of the great stone barn with the loathing of emigration? The explanation belongs to the meaning of the hay husbandry to northern improvers and the way they thought about the West. They never laid their bodies across the road to oppose continental expansion; instead, they stood by with queasy uncertainty. Did it represent the rise of a new civilization or the twilight of an older one? Most of all, they disdained frontiering—that untidy process of Indian-like burning and retrograde settlement that they assumed could not possibly bring about social stability. Their own neighborhoods had long since emerged from that crude epoch, yet emigration suggested that they might still suffer its privations in the form of lost laborers and in the rise of new states unbeholden to the old. Improvement did not simply offer to boost yields and profits. It proposed a hierarchy of land use in which rich sod represented the maturity of American civilization. Disdain for the wasteful ways of backwoods people became a negative source of identity for the self-conscious people who stayed behind. The high husbandry became the outward sign of their salvation.

James Pemberton Morris felt the unease of one who remained behind. Addressing the Agricultural Society of Bucks County in the middle of a busy April season in 1823, Morris filled in the missing matter between the confident farmer in a good country and the disquieted elite. He advocated education as a ramp to lift husbandry to the social status of the professions. "Agriculture would then become the resort of the rich, and the refined classes of the community, and would no longer be considered as the portion only of those, whose classification is among clowns." Farmers with diplomas would make country life an honorable retreat for survivors of hard times. The wealthy merchant and attorney would be comforted in retirement knowing that they had secured for their children "a succession that shall be safe from . . . . the fickleness of fortune," without banks or brokers, "and none shall make them afraid." Morris's way of thinking smoothed out the likeness between agrarian republicans and soft-handed burghers. His hybrid political identity anchored him in the soil of agrarian self-sufficiency but still allowed him to consort with banks and brokers and float the muddy stream of paper wealth. The self-improvement first made popular by William Ellery Channing, a Unitarian minister, seems fitting here: "Improve, then, your lot. Multiply comforts, and still more, get wealth, if you can, by honorable means." In Morris's idyll, the improved countryside made gentlemen out of farmers and farmers out of gentlemen.[33]

Morris also spoke about the farm he hoped to leave behind, another dimension of his old-state identity:

> The trees which we have planted with our own hands in infancy and in youth, are those under whose shades we will receive the greatest solace in declining age. The flowers and the fruits which have regaled the morning of our lives, will retain, in its evening hours, all their fragrance and all their flavour. These associations of an early education, have an influence almost unknown to ourselves, even in preserving and continuing our farms to our descendants. They soon become, when thus associated, a sacred patrimony, which we would while living defend with our lives, and when dying our most fervent prayer would be, that they might descend to our children.[34]

Why bless permanence in so public and personal a way unless it needed to be affirmed? Morris sensed that permanence had declined in the mind of the public, like a neglected virtue. He feared being forgotten where he lived, shuddered at the thought that his life's work might one day be sold to strangers. And the people who forsook the "sacred patrimony" of land? He mentions clowns more than once, at times to mean all farmers unfairly categorized, at other times to mean a crude class of settler.[35] The image of clownish people at city gates, funny in dress, awkward in speech, is an urban conceit as old as Athens. Morris might have meant to say that all farmers deserved respect from the people in town whom they fed and clothed, but what he was really doing was seizing that respect for himself by drawing distinctions between farmers. The clown, in this other sense, represented the yeoman who never moved beyond first-generation tillage or backwoods manners: not the people who stayed, joined the society, and built great barns, but the others. Another resident of Pennsylvania had more to say on the subject.

Benjamin Rush, member of the Continental Congress, doctor, and professor of medicine at the College of Pennsylvania, published "An Account of the Progress of Population, Agriculture, Manners, and Government in Pennsylvania," which was filled with criticism for frontier farming and depicted the ascent of civilization. "The first settler in the woods is generally a man who has outlived his credit or fortune in the cultivated parts of the State," sniffed Rush. Negativity so utterly uncurbed reveals all the positive qualities that Rush associated with the older estates: The yeoman "lives in the neighbourhood of Indians . . . soon acquires a strong tincture of their manners"; "loves spirituous liquors, and he eats, drinks and sleeps in dirt and rags"; "above all, he revolts against the operation of laws." Any worthy improvements to property came from the second class of settler. Number two tried to do better but fumbled everything, incompetent in cultivation. Watch him plow too little and not deeply; see his cattle break through the half-made fence and graze the grain. He diversified his arable to include crops other than Indian corn and might even have built a log barn to shelter his animals. He had a house, though not a very dignified one (unglazed windows), all of which exhibited "a weak tone of mind." No church here, or any of the other institutions of civil government; this settler, like the first, hated taxes and loved liberty too much.

For Rush, the third "species of settler" arrives clean and well equipped, the son of a wealthy farmer from one of the "ancient counties," a man of good character. Know the species by its habitat: big stone barns and green sod. "His first object is to convert every spot of ground, over which he is able to draw water, into meadow." Most important for Benjamin Rush, "it is in the third species of settlers only, that we behold civilization completed—it is to the third species of settlers only, that it is proper to apply the term of farmers." Rush narrated a complete cycle of social evolution—"progress from the savage to the civilized life"—through the adoption of specific agricultural practices. Unsure of the size or extent of the backwoods frontier stretching out from under his feet, Rush demarcated civilization from barbarism with a line that just happened to follow the gentle curve of the Allegheny Front. He strongly implied on which side authority should reside—with the people who created permanence and held it as a virtue. For all their lowliness, frontier farmers, Rush recognized, served a grand historical purpose. They represented a crucial evolutionary stage in the progress of civilization in which their removal made way for the "frugal and industrious." In the 1790s, Rush's third farmer was a distinct minority in the counties of southeastern Pennsylvania and Delaware, but less so by the 1820s.[36]

Smugness alternated with raving, paranoid anger. Sometimes the cattle show became the place where easterners vented their fears about national expansion. Said one speaker at Worcester, Massachusetts, bent on prophecy in 1824: "Let our literary institutions decay . . . and but a few generations will pass before our descendants from the Hesperian gardens of the West will look in vain for the neat School-House, for the Temples of Religion. . . . The traveler may indeed find a bloated population—he may find fields of rank luxuriance—he may hear the noise of the Mechanic, and see the busy stir of commerce; but it will be a 'barren splendour,' a 'sickly greatness,' a 'florid vigour,' betokening disease, decay and death." The poor emigrant becomes dominated by "a few supercilious lordlings." Spies, thieves, robbers, and assassins—corruption of all kinds flows from the West like high waters, soon to "desolate the land" and sweep away the rule of law. "Discontent and misery" set in among the people, who, after "listening to a succession of demagogues, will be prepared to surrender the remnant of their liberties to the iron grasp of a Dictator."[37] In the aftermath of the Burr conspiracy it seemed possible that a region in reckless isolation might break away, drunk and disorderly with its own rank luxuriance.

Bouts of negativity do not capture the ideas at work. Improvement in the North cannot be understood as the mere complaint of people accustomed to wielding social status or as frustrated backlash to events beyond their control. The literature is replete with obligation—not just duty to state and nation but moral duty to take good care, to leave the ground as full as first broken. This sentiment, detached (at least in rhetoric) from any obvious interest, characterized northern improvers in what was, in essence, a proto-Whig attitude toward nature. The president of the Pennsylvania Agricultural Society pronounced the ethic in an address of 1823: "The history of all times proves that moral advancements are the more durable for having been gradually attained. Their solidity, it would seem, was in the ratio of their march." He then tied this sober ideal directly to the sufferings of land: "In many of the settlements around us, the natural fertility of the soil has been exhausted. Comparatively little new grounds remain to be brought under cultivation. To refertilize the old must be our resource."[38] This "organic conservatism" regarded society as a cell in which individuals derived identity from established institutions—churches, town meetings, political clubs. Community embodied tradition anchored in place and needed immunity against a host of threats to its integrity, because when organelles breach the cell wall the nucleus dies.[39] It was a view more popular in the South than in the North (recall John Taylor's nation by nature), but it is evident in these sources. The commissioner of the Massachusetts Agricultural Survey hollered out over the cattle show at Worcester in 1838: "Farmers! To your country you owe high duties. . . . You are fixtures to the soil. Other men may at pleasure transfer their residence, interests, and affections. It is not so with you. Your interests and fortunes are indissolubly entwined with the interests and fortunes of your country."[40] It was the nucleus talking.

Conservation is barely distinguishable from conservatism in these examples, according to their common root: to preserve from decay or loss, to maintain in continuous existence. The most articulate representative of this sentiment was Jesse Buel, editor of the *Cultivator*. Buel was born in 1778 in Coventry, Connecticut, and his family moved to Rutland, Vermont, in 1790. There Buel apprenticed to a printer and married. He then managed a number of weekly newspapers in Troy and Poughkeepsie, including the anti-Federalist *Ulster Plebeian*. Buel founded the Albany *Argus* in 1813 but gave it up in 1821, when he purchased eighty-five acres of the "Sandy Barrens" west of Albany to establish a farm. "Bred to a mechanical business," in his words, Buel took up agriculture "from choice, as the future business of my life." During the 1820s he emerged as a self-educated tiller, a founder of the Rensselaer Polytechnic Institute, and a Whig politician. He established the *Cultivator* in 1834. Two years later he made a run for governor but lost to William L. Marcy. In the 1830s

Buel came out with a number of important books and essays, each of which asserted ethical arguments for the best possible care of farmland.[41] "It is your province, and your duty, to husband and apply the vegetable, and most essential element of fertility—MANURES. . . . It results from these facts that a farmer should till no more land than he can keep dry, and clean, and rich; and that he should keep no more stock than his crops will feed well, and that can be made profitable to the farm." In another statement, exemplary of the eloquence that permanence inspired, Buel writes about land as an earthly Constitution:

> The new system of husbandry is based upon the belief, that our lands will not wear out, or become exhausted of their fertility, if they are judiciously managed; but, on the contrary, that they may be made progressively to increase in product,—in rewards to the husbandman, and in benefits to society, at least for some time to come. It regards the soil as a gift of the beneficent Creator, in which we hold but a life estate, and which, like our free institutions, we are bound to transmit, UNIMPAIRED, to posterity.[42]

Responsibility, self-denial, ties to society through family, and the solemn transmittal of values and practices; social mobility by hard work, not greed; social order and leadership by respected elites, not demagogues; education, religion, and temperance in all things—these describe the convictions of the people who became Whigs in the 1830s. Buel's words illustrate virtually every facet of Whiggery, from his constitutional language (forming a parallel between the order of private property and that of the Republic) to his belief that social and moral progress depended on economic progress. Whigs held that freedom never came without duty, and betterment never without hard and meaningful work.[43] Buel brought this political philosophy to the furrow slice: an enduring cultivation represented the best social principles as well as that habit of discipline that made for material and moral improvement. The farm needed proper comportment to thrive, and nature needed the civility of cultivation in a parallel with the human soul.

When, some years later, George Perkins Marsh wrote about the neglected commons in newly settled countries, he participated in the same tradition: "This decay should be arrested, and . . . the future operations of rural husbandry and of forest industry . . . should be so conducted as to prevent the widespread mischiefs which have been elsewhere produced by thoughtless or wanton destruction of the natural safeguards of the soil." After all, Marsh had been shaped by the same environment. When Buel founded the *Cultivator* in the city of Albany in 1834 at the age of fifty-six, Marsh was thirty-three, a lawyer in Burlington, Vermont, the partial owner of a seven-story woolens mill, and soon to be a Whig candidate for Congress.[44] Marsh came of age during the peak intensity of improvement in the North. Beginning in the 1840s, he intellectualized the painful changes he had seen where he grew up and married an early affinity for watersheds with his enormous talent as a scholar. The same high-minded concern of the Philadelphia gentlemen emerged from the pen of an imaginative thinker as a vigorous philosophy of resource consumption.

The philosophy of permanence created a countryside of exclusion, in sharp contrast to the classless society of the frontier. The members of the Philadelphia Society for Promoting Agriculture never seemed to have understood that southeastern Pennsylvania had simply filled up and become more costly to cultivate. Not everyone could afford to buy a farm rich enough to yield thirty bushels per acre. Not everyone could afford the big barn and the dung pile. Emigration represented a predictable response to density, like the flow of any substance from a higher to a lower concentration. At the end of the eighteenth century at least one-third of the young men under the age of thirty-five in New England and the Middle Atlantic states worked as landless laborers while their parents grew older and more dependent on the income of the family farm. An English tradesman, his eyes and ears attentive to the conditions of labor in New York, noticed plenty of casual workers but a scarcity of land: "Farmers here have no difficulty in getting labourers. . . . In this and similar districts a poor man has little chance of buying land." Benjamin Rush never imagined the debt he owed to the people he dismissed as the scruffy undergrowth of society whose dislocation he cheered for opening the ground to an unfailing cultivation. Poor farmers and their landless offspring created the landscape that Rush valorized; the good husbandry depended on the same high property values that caused young families to emigrate. John Lorain even asserted that convertible husbandry could not be practiced at all below an unspecified threshold of population: "It cannot be generally practiced even in the populous parts of this country. The quantity of cleared ground is more than double as much as the population is capable of cultivating properly, without introducing the additional labor which would be required if soiling were generally practiced." Density served the ends of improvement.[45]

In other words, an ever higher population pressing on a limited extent of land resulted in landlessness. Southern planters thought differently about landlessness. Having solved their own labor problem only by enslaving African-Americans, they became champions of republican opportunity for whites. Though perfect unity never existed among them, generally speaking, planters advocated the unfettered disposal of the public lands, especially when that meant the formation of new slave states. First among land-loving planters, Thomas Jefferson translated the work of the French economist Destutt de Tracy, who warned that a Malthusian force of numbers would eventually overtake the natural endowment of any nation. The time would come when all the good land would be held by rich proprietors, leaving the great majority to fend in a population of dirt-poor wageworkers. "They quickly run into debt," he said of the indigent, "and are necessarily turned away. Yet others are always found to replace them, because [there] are always wretched people who know not what to do."[46] Jefferson and his followers recognized in that scenario the destruction of the republican experiment.

Here was a fundamental disagreement between the sections based on the uses of land, its availability, and the legal structures surrounding it. For southerners committed to republicanism, western lands made the continuance of yeoman settlement possible and held off some very big problems: white landlessness, meddlesome tariffs, and a northern-dominated Congress. For northerners insisting on their cultural leadership, seething with disdain for the "sickly greatness" of the West, blank spaces on the map represented the dissipation and debasement of the American nation, not its golden future. The disagreement finally took the form of a short resolution in the United States Senate, introduced by Samuel Foot of Connecticut during the first session of the Twenty-first Congress, in 1829. Although the Foot Resolution had nothing to do with the mucky particulars of husbandry, it was the last time that a coterie of New Englanders attempted to regulate frontier settlement and use statute power to throw up a wall around the old thirteen.

The Foot Resolution specified that Congress limit the sale of public lands to those acres already surveyed and abolish the office of surveyor general. Yet those few words contained an entire vision of national progress and incited a debate about nullification, sectionalism, and federal power. The gesticulating went on for four months, with senators rising to speak for nearly days at a time. Daniel Webster delivered his famous "Reply to Hayne," in which he pronounced the argument, picked up by Abraham Lincoln, that the Constitution formed a compact with the people, not with the states, and created a Union of mystical authority, inviolate in all cases, "now and forever." By proposing an artificial limit on sales, the Foot Resolution pitted permanence against expansion on the highest stage of public display. Much of the motive lies in the life of the author. Samuel Augustus Foot returned home to Cheshire, Connecticut, following law school and a start in the shipping business. He operated a farm for four years, beginning in 1813. Then came his political career. Standing up for John Quincy Adams against Andrew Jackson in 1828, and then as a Whig candidate for governor in 1836, Foot embodied eastern conservatism, with its emphasis on the obligations of property and its fear of popular democracy. Although he and other members of the New England delegation repeatedly claimed that they had no intention of undermining western progress, they fully intended to slow it.[47]

The debate played out along sectional lines. Thomas Hart Benton spent most of his career in the Senate trying to lower the price of western lands, so the Foot Resolution made him churn with anger at

New England: "This resolution . . . is the true measure for supplying the poor people which the manufactories need. It proposes to take away the inducement to emigration. It takes all of the fresh lands out of market. It . . . annihilates the very object of attraction—breaks and destroys the [magnet] which was drawing the people of the Northeast to the blooming regions of the West." Benton had said it all before in 1828, when he accused the secretary of the Treasury, Richard Rush (son of Benjamin Rush), of holding the price of land above value "to check the growth and prosperity of that part of the Union for the sake of promoting the prosperity of another."[48] Indignant but hardly speechless, the New England delegation attempted to articulate a policy of restraint, slow growth, and federal paternalism over grab and gain. Supporters of the resolution wondered why the West and South resisted a temporary ban on new surveys so that the present population could take up the tens of millions of acres already set aside. The next order of business during the same session would be the violent removal of the southeastern tribes to the far side of the Mississippi River, causing unimaginable suffering, and for what? . . . to open yet more land. In the few instances where they actually spoke in favor of the resolution (and not simply contra the arguments offered by their opponents), New Englanders worried that expansion led to waste. Give one turkey to each of twenty dinner guests, said John Holmes of Maine, "and they are all left partially eaten and all mangled. Now this is exactly the case of bringing more land into the market than can possibly be wanted." Distaste for waste is what made the resolution kin to improvement.

Southern radicals spoke through Robert Hayne of South Carolina and made common cause with the West just to bloody New England for the dreaded Tariff Act of 1828. The tariff protected northern farmers and manufacturers against foreign imports of wool and woolens. It seemed to hatch from a cabal of northern interests indifferent to the needs of planters. The apparent dictatorship of one section over the other made for an especially volatile fallout. The tariff spread John Taylor's old rage against bounties throughout the South and inspired John C. Calhoun to put down a doctrine asserting the right of a state to nullify any federal law detrimental to its interests.

Webster's famous reply to Hayne was not his first—that had come a week before. In it, he insisted that New England had always supported measures for western development—like canals and the Cumberland Road—even at its own expense. His section endeavored to advance the national good, while the others engaged in divisive politics, threatening "nullification" to feather their own beds.[49] Southern radicals found an ally in Benton, for whom nullification offered a constitutional vehicle to extinguish federal authority over the public lands in Missouri.

It all came to a head when Hayne lined up against the resolution, even though his own state almost never supported internal improvement in the West, even though western settlement in no way advanced the interests of South Carolina. Southerners may have supported the public domain because it provided the raw material for an expanding nation of slave owners, but they also dreaded its attractions. Leading planters had been sounding a fire bell about the accelerating pace of emigration since 1818. Even if Hayne counted himself a western booster, he must have understood that South Carolina only stood to lose from expansion. Realizing this, Webster brought the bad consequences of emigration into the debate and embarrassed Hayne by quoting from the remarks of George McDuffie, a representative from one of South Carolina's Piedmont districts whom Webster had debated in 1825 on the issue of the western road. Said McDuffie back then, " 'Deserted villages, houses falling into ruin, impoverished lands thrown out of cultivation. Sir, I believe that, if the public lands had never been sold, the aggregate amount of the national wealth would have been greater at this moment. Our population, if concentrated in the old States, and not ground down by tariffs, would have been more prosperous and more wealthy.' "[50] With those remarks, Webster threw cold water on the West-South alliance, revealing its contradictory interests, but the debate bore nothing. After popping up again in May, the matter died of a motion to set it on the table, so we have no vote to mull over.

Never mind the intrigues of the Senate. Emigration had become politics at the highest level and came to represent, in the minds of western and southern senators, the demise of an entrenched power in the North and the birth of a more expansive sovereignty in which slavery might broaden and spread, in which industry might matter less, in which the extensive use of land would equal political power to fly in the face of the people with barns made of stone. The Foot Resolution attempted to manage the wild and politically unpredictable space called the public domain and to encourage a greater density of settlement and a more responsible and traditional regard for the value of property. But though their cultural engineering failed, their land-use philosophy remained. Government regulation in an era of ascending democracy was destined for tatters, but persuasion might still build an eastern garden. Samuel Foot and his allies would have agreed with George Marsh when he reflected, some years later, on the diminished cultural authority of New England next to a surging and insubordinate western frontier: "If then we cannot be the legislators of our common country, let it be your care that we become not unworthy to be its teachers, and though we cannot give it law, let us not cease to give it light." One capable farmer ventured close to where he could observe ax-and-rifle settlers, and there he hoped to teach them.

Hints to Emigrants

To those unsympathetic to their yearning—including just about every Briton who wrote about them—emigrants suffered for no reason. In such accounts they endanger the lives of their children, starve on the road, live filthily. Henry Bradshaw Fearon, an English traveler following the human tide, came to the Juniata River and, after crossing it on the western road at the town of Bedford, continued up Dry Ridge, where he saw families bound for Illinois. One woman out thirty-two days told him, "Ah! Sir, I wish to God we had never left home." From Dry Ridge to the Blue Ridge he saw overlanders formerly of Maine, Maryland, Pennsylvania, Connecticut, Massachusetts, and New Jersey, many of them "in great distress."[51]

John Lorain saw them too. High in Centre County, near the present route of Interstate 80, he kept track of the back-and-forth: "That many [have] returned from the western country is as well known to those living on the roads leading to and from it, as any other fact." Many others would have retraced their ruts from "this land of sickness and premature death" had they not blown all their money to get there.[52] Far more interesting to Lorain and more telling to him about the success or failure of emigration were the practices backwoods people put to use once they arrived. Lots of people claimed to know the taste of tilth in the 1820s, but in a time of contrary conclusions and undependable experiments Lorain spoke with placid authority. He could girdle a tree as well as soil cattle. In sharp departure from others of his class and education, he sometimes placed the frontier farmer and the book-learned gentleman on an equal footing, or hauled them in for equal reproach. He could see both in terms of their desires and social settings. Lorain attempted to translate the restorative husbandry of English theory across borders of class and geography and to resolve the dichotomy between the Philadelphia Society and the dynamic settler culture pounding out the shape of a new world beyond Pittsburgh. He published a string of essays in the society's *Memoirs,* and the members clearly valued his work. But he communicated with them at a distance. John Lorain did not live in Philadelphia or in Bucks County or even in the foothills. He lived near the mountains, having moved there, indeed emigrated there, in his maturity, and that made him different. In his able hands, a steady cultivation became the foundation for a moral and economic critique of the West.

Not a letter under his name survives, and the little I have discovered about his life can be reported in a paragraph. John Lorain was born on the Eastern Shore of Maryland in 1753, soon after his parents arrived from England. There he established himself as a planter and slave owner until the age of forty-two, when he moved to Germantown, near Philadelphia. He joined the Philadelphia Society for Promoting Agriculture in 1810 and moved again to take up residence in that city. A directory of 1811 lists his address as 55 North Front Street, but by 1814 his name has disappeared from those lists. He is said to have answered an invitation from a friend to come to Philipsburg, and apparently made the move in 1813. He may have purchased the Simler property on North Second Street and opened a general store. The citizens made him postmaster and then justice of the peace soon after his arrival. And by his own account, he and his wife, Martha, lived in a log cabin not far from town, on a farm settled and then abandoned by one of the first settlers. There he assumed the life of a plain practical farmer, with no interest in the fluff and ostentation that attracted city gentlemen to husbandry, revealing in one instance that, "of the cattle show, I know little, having never been there." He published one book and many articles before his death on July 22, 1823. A second book, *Nature and Reason*, went to press posthumously the next year, with Martha's critical intervention.[53]

Philipsburg sits in the middle of Pennsylvania, on Moshannon Creek, right where the ridge-and-valley region meets the bold escarpment called the Allegheny Front, where agriculture stops cold.[54] A map of 1822 indicates iron and coal in the region. The yellow clay deposits in the creek made good brick. The first screw mill in the United States went into business there. Philipsburg, founded circa 1797, probably counted fewer people than the next major town, Bellefonte, which had a population of 433 in 1820. The entire county claimed only 13,796 citizens in the same year. For comparison, Pittsburgh had 7,248 and its county, Allegheny, had 35,000. One report held that "in the counties of Centre, Bedford, and Huntington, many of the inhabitants live by hunting and fishing, and gathering wild honey from the hollow trunks of trees"—in other words, they lived like first settlers.[55] Why would someone welling with criticism for the West move from the courteous environs of Philadelphia to backwater Philipsburg not fifteen years after its founders hacked it from a mountain clearing? Why would a farmer of judgment, perhaps even genius, move to a place of such sparseness, which, according to one foreign observer, utterly lacked the advantages of a good settlers' country?

John Pearson, another journalist recently arrived from England, also documented the emigration of the 1820s and wrote one of the only descriptions of Centre County. He called the place rocky and barren, and treacherous in other ways as well:

> We now directed our steps towards Philipsburgh . . . the land is mostly covered with stones, and the greater part is pine, which

> is the worst of land . . . it is in the very heart of a large wilderness, far removed from towns or villages, consequently no change of society, without which life itself becomes a burden; it is also [in] an infant state, and will of course require vast sums of money and labour to complete what is intended by Mr. Phillips and his shoals of sharks and land jobbing friends, whom I consider in no other light than a gang of kidnappers and plunderers, working upon the credulity of Englishmen.[56]

Pearson was convinced that boosters of this snaggletooth mountain town set out to separate unsuspecting foreigners from their money by selling them sterile land and a rude, miserable existence. At one point the traveler encountered an elderly man preparing for a trip down the Ohio River. "I told him he had better stay where he was, if he knew when he was well off, for I thought he stood more in need of a coffin at his time of life, than to rattle his poor old bones in a wild goose chase down the Ohio, unless it was his intention to join Messers. Johnson, Lorain, Philips, Rose, Jennings, Birkbeck, and Co. and so work the gold mines with them, or manufacture the philosopher's stone." Pearson had seen emigrants ripped off all along his journey and associated these names with the crime. Did Lorain practice this art, as Pearson asserted?

Controversy over the western lands is the perfect entrance to a picture of Lorain's work, and that subject will always be tied to another name on Pearson's list of villains: Morris Birkbeck. Pearson only hinted at the bitterness that Birkbeck and his Illinois settlement provoked when he asked the misguided man on his way down the Ohio to "give my kind respects to [Mr. Birkbeck], and tell him that he had met with a man who was now 150 guineas the worse for his letters and notes."[57] To have known of Birkbeck's letters and notes was to have had a strong opinion of them, and Lorain knew of them. After the panic and into the 1820s, only a few farmers attacked the Northwest as a lie of abundance. Lorain's complaint had less to do with duplicity and sharp dealing than with the true prospects and future happiness of the emigrants. No one else in the country spoke like the improvers; no one else stood against the expansion of population into the territories; no one else articulated a principle of measured growth.

Morris Birkbeck, formerly an English landlord, published a series of letters from Illinois intended to advertise its ease of settlement and many advantages to the humble farmers of his native home, whom he expected would cross the Atlantic and find trails to the prairie, where they would purchase land from him. Birkbeck's book made a bigger splash than many other works of its kind and remained popular for years after it appeared in 1818 (I read from the copy once owned by the Mormon founder and prophet, Joseph Smith, who might have used it to scout out the settlement at Nauvoo in the 1830s). What so angered Lorain, aside from the book's appalling lies ("A single settler may get his labour done by the piece on moderate terms, not higher than in some parts of England"), was its glowing claims about the country, especially that the seeking farmer would discover land that needed no manure, with "no bounds to its fertility." More than that, Birkbeck misrepresented what the average settler could achieve in Illinois by proposing an estate for himself to rival Wanborough Farm, the seat he left behind. In other words, Birkbeck set out to shape 1,400 acres of prairie (plus 160 acres of timber) into the same country house, outbuildings, five thousand rods of fence, gristmill, and labor force that he once knew. Birkbeck did not deny that settlers suffered, but all privation is quickly overcome in his account, and he told emigrants to expect comfort and ease soon after arrival: "To have passed through all this harmless, and even triumphantly, to have secured a retreat for ourselves . . . is an enjoyment which I did not calculate upon when we quitted our old home."[58]

Birkbeck committed every offense against Lorain's sensibility: land gluttony with aristocratic airs, overlaid with speculative greed. But more than argue against Birkbeck, Lorain argued in favor of old ground. He never asserted any specific social or political objective, but by defending his place and process, he recommended another philosophy. The kind of cultivation Lorain advocated could be conducted with free and scarce labor, on small acres, resulting in middling prosperity, or in what Daniel Defoe's Robinson Crusoe called that "high station of low life," promising the least worry. He seems to have moved such a great distance to make a point—that improved husbandry could be profitably applied to western lands—and he took possession of a skimmed-over farmstead to prove it. True, Lorain's clearing was not so old, but he bought it worn-out by its previous owner, so it represented the very land that frontier settlers typically abandoned after fifteen or twenty years. He used it to draw a line. If this thin and stony land, of a poor endowment, rife with pine, "the worst of land," could be made as productive as virgin prairie, then no farmer ever had any reason to emigrate for fertility:

> I have grown much more than twenty bushels of wheat to the acre, on the hill and dale lands laying a mile and a half from Philipsburg immediately after maize, the wheat sown while the corn plants were still standing on the ground. The maize also yielded much more to the acre than Mr. Birkbeck estimates as the produce of his prairie; and it was grown on land which had been considerably exhausted by very severe cropping, with but little attention to grass or manure.

This was the core of the matter for Lorain. Farmers who deserted the Atlantic states for the flats looking for a magic soil were badly mistaken. If the Illinois settler had instead "settled on any of the hill and dale lands, in the interior or back parts of the middle states, where the soil was only moderately rich, and given the same attention to manure, with due regard to grass, he would, on an average of years, have gathered more produce from the same breadth of soil, and also have sold it at prices more than exceeding double, those he will be able to obtain in his present situation."[59] Pearson, the accusing English traveler, probably thought that Lorain advertised Centre County and thundered against Birkbeck's prairie for his own benefit, but Lorain only used himself as an example of modest prosperity to encourage persistence in others.

Lorain made these arguments in *Hints to Emigrants*, a tract he aimed directly at Birkbeck's Illinois speculation. But the book was not the only or the most notable of the denunciations against Birkbeck. William Cobbett, an English political writer who spent a year in the United States and came to know Pennsylvania, wrote his own fuming letter: "It is the Prairie, that pretty French word, which means green grass bespangled with daisies and cowslips! Oh God! What delusion!" Cobbett challenged the very idea of removal for material advantage: "Before we proceed to enquire, whether such persons ought to emigrate to the West or the East, it may not be amiss to enquire a little, whether they ought to emigrate at all! Do not start, now! For, while I am very certain that emigration of such persons is not, in the end, calculated to produce benefits to America, as a nation, I greatly doubt of its being, generally speaking, of any benefit to the emigrants themselves." Cobbett expressed the ethic of land-conservation-as-self-regulation so attractive to North American improvers. Just because settlers could move to the new state of Illinois did not mean that they should. The two Englishmen knew each other and spoke in London before Birkbeck's departure. Said Cobbett of the meeting: "I begged of you . . . not to think of the Wilderness. I begged of you to go to within a day's ride of some of these great cities, where your ample capital and your great skill could not fail to place you upon a footing, at least, with the richest amongst the most happy and enlightened Yeomanry in the world." "The truth is," said Cobbett, finishing off the subject in one breath, "that this is not transplanting, it is tearing up and flinging away."[60]

The frontier, however, had already captivated the rural public, many of whom did not need Birkbeck's letters to make their blood run fast at the thought of so much land so good to break, and Lorain knew it. Philipsburg served as a perch from where he could see emigrant trails and smell the smoke of expansion. He watched wilderness turn to plowland and wrote about that process like one who had done it himself. As a former resident of Pennsylvania's southeast and as a member of the Philadelphia Society for Promoting Agriculture, he also knew the culture of rural affluence and romanticism. An arbiter of the wildly contradictory information in much agricultural discourse, Lorain put down his most detailed and comprehensive thinking about cultivation in *Nature and Reason Harmonized in the Practice of Husbandry*, a major treatise on American land use in the 1820s. In this work he accomplished what no other writer at the time even attempted: he understood the opposite methods and motivations of gentlemen and skimmers and proposed a kind of agronomic synthesis. Lorain studied the physiology of settled society, knew its biological systems and how people used them for better or worse. Backwoods farmers are not stupid in his book, and gentlemen are not very smart. They simply live on different sides of a rural economy, leading to very different tenures of land.

"Since I removed to the back-woods, where [frontier] farmers are more especially plenty, I have had a better opportunity of observing their talents, and must confess that I have found them much more respectable than I formerly believed them to be." Any one of them could build an ark without iron or cordage and pilot it in furious freshets, even weighted with cargo; they walked miles without complaint to provide necessities for their families, and as travelers they became informed about watercourses and learned how to barter and purchase. They framed houses with nothing but an ax, an auger, and a knife—good doors, strong floors, and furniture constructed without embellishment and without a single nail. They came "exceedingly poor," better stocked with children than with cattle, so Lorain learned to admire them. Most of all he wanted to inform them, but in order to do that he first had to figure out what they needed to know and what they could possibly afford. Others of like mind ran on for pages with the former but choked on the latter. Lorain's first premise was that the high husbandry of England contained many flaws. Its core practices assumed limitless capital. He called it "arduous, especially when ventured upon by a plain practical farmer, who depends on science, but [also] on nature, reason, practice, and observation," and even recommended that the same big, gorgeous muck-making system that worked so well near the seaboard not be practiced except "where population is thick, and labour easily obtained."[61] A democratic cultivation would take the best and leave the rest for the nodding adherents of Arthur Young.

Yet Lorain had no illusions about his neighbors. When it came to land, they did most things badly, or, in other words, "there are no cultivators of the soil in Pennsylvania, that farm worse than do these men." He also believed that given their energy for the hardest work, backwoods settlers could farm as well as any other people in the world "who possess no more capital than they do." He did not abandon them. Instead, he considered every minute particular of their cultivation to direct it in the most profitable direction. He followed highbrow convention by denouncing all burning as "savage," claiming that humus gave up its nutrients better without it (even though fire lingered for a while in English practice and in spite of the fact that poor farmers all over the world used it as the most labor-efficient way of reducing biomass). Nonetheless, rather than reject the pioneer use of fire and turn his back, Lorain advised on the subject: burn during a wet season to reduce the leaves and dry matter without harming the thick humus underneath; or better yet, rake that debris into piles, burn it, and then mix it in after plowing.[62] He wanted to understand and influence how backwoods people shaped land for subsistence.

Lorain lived on the cusp of expansion, in the very hearth where geographers believe the American backwoods culture began. The people crossing the Appalachian Front near his door would settle more land more quickly than any others in human history and would, in less than twenty years, end up in Oregon. To do this, they needed more than cattle and plows. Their subsistence included meat from hunting and fruits from gathering, each promising fairly abundant calories for little labor. Lorain had nothing to say about skills for living in great wide-open territory, probably for the same reasons that Benjamin Rush condemned them. They made white settlers resemble Indians. More striking for a farmer, he had nothing to say about swine. Pigs followed the diffusion of backwoods culture because they created flesh from feed in a tight ratio and propagated on the barest subsistence. But pigs did not figure in theories of landscape stability, so he ignored them. It was a prejudice he shared with the gentlemen of Philadelphia. Like them, Lorain feared western settlers, especially their scorching ambition for "immediate advantage over eventual gain." A cultivation that depended on unbroken prairie had the potential for stunning ecological change, and in his writings he constantly tried to contain that force. "Man," he wrote abruptly, "is the most destructive animal in the universe, when he considers that his resources cannot fail."[63]

Destructive, and sometimes ridiculous. While Dr. Rush considered backwoods people stuck in a Neanderthal moment of evolution, John Lorain would have informed him that gentlemen also failed at agriculture. They used the best English theory but followed it blindly. In their uncallused hands, the latest methods paid little interest and promoted little more than "extravagant parade and show." Lorain never doubted fallow crops and winter penning, but it clearly frustrated him to see them adopted without attention to economy or environment. In a word, the puffy esquire who went to the countryside with visions of Holkham often lacked the common sense of a woodsman. Gentlemen had no idea what their crops were worth and expected far too much at market, as though husbandry were a snap, as though profits would pile up on the floor like the backside of a cow. Lorain said that the affluent appeared to forget or not to know "that agriculture, when properly pursued, under the most favourable circumstances, requires very great attention, both early and late; and that there are very few employments which have more crosses, losses and disappointments." Common farmers never made these kinds of mistakes. "On the contrary, they show their good sense by not giving into practices that would infallibly ruin them."[64] The readers of theory needed help no less than western emigrants, because they also sustained families and communities through the changes they made to land.

Nothing better symbolized the misplaced efforts of esquires than the houses they built—all fuss and trimming, "a botched piece of business" totally out of proportion with what the farm would support—making the entire enterprise not farming but something else, a preening of feathers for public consumption. Their purchases proved their ineptitude: barns with marble ornament; fancy fruit trees where ordinary ones stood before; every imaginable implement of husbandry first read about in books, now given to workmen to figure out; the finest breeds for dairy and work animals, though the owner has no idea how to keep the animals, though his wife refuses to squat low on the animal, "over her shoe tops in mud or dung, or half leg deep in snow, until the cows are milked." The yard is renamed the lawn. The gentleman hardly noticed when he condemned an old farmhouse to build his picturesque residences that thousands suffering economic depression and wishing nothing more than plain comfort "were continually leaving the country, and flocking into our cities and towns, with as much avidity as if happiness were no where else to be found."[65] When he gets tired of farming and goes to sell, "as too commonly happens," he finds that all the money he spent over and above basic improvements is lost. "Useless brick, mortar and lumber, united with every expensive ornamental work, form no part of the estimate made of the value of the premises by a prudent purchaser. He knows full well, that no profit which can be derived from farming, will pay him an interest on it."[66] The beauty of it is that Lorain instructed the feckless rich with the same attention, the same grace, and the same sympathy he gave the backwoods poor. Information that any common farmer would have found elementary—figure things per acre, buy no livestock until you remove to the farm—amounted to a kind of remedial course of study. Lorain even advised on a will. Holders of country seats should separate buildings from cropland for the beneficiaries of the estate. The gentleman would more likely feel "easy on his dying pillow" knowing that the farm was free and clear, with an annual income unencumbered by the cost and overvaluation of the house. Pretending to live as a farmer, the squire might as well die with the satisfaction of one.

Aristotle would have recognized Lorain's logic, for between an excess and a deficiency of any quality there lies the Mean—the proper apportionment for right action. Lorain finally presents a middle way of his own between skimming and preening as a responsible and productive American cultivation, flexible enough to maintain the East and steady the West. The problem of the two regions was really one and the same because, as he said, there is no difference between farmers who cannot and those who do not apply the right capital and proper knowledge to the care of land. The middle way, called the method for "circumscribed farmers," illustrates both financial and environmental economy, a cultivation for both sides of the ridge. No surprise, the solution began with grass, "nature's infallible restorative." If poor farmers would do nothing more than plant quality grasses (red clover and timothy), they would discover a way to hold their soil, increase the health and number of their stock, and better their land and themselves. Manure should be collected and spread over a small planting of peas and beans, and the acres dedicated to these crops should be extended as the number of livestock increased until the fallow, the grass, and the arable reach that critical compression that ignites into "one perfect system of economy."

Grass formed only one part. The new settler should avoid plowing and live for as long as possible off animal products—milk, butter, cheese, wool, meat. These increase all by themselves even with wild forage. "From first to last," Lorain told readers, those who have kept off plowing while their stock increased "have been enabled to live better, and vastly more independently than those who relied principally on the plough." The complete system could wait for stability. Lorain's stripped-down version of the Norfolk system aimed to "promote the ease and tranquillity of an indigent cultivator, and finally to enrich him."[67] Yet Lorain seems never to have appreciated the irony that he attempted to teach the methods of intensive, commercial agriculture to yeoman farmers, who often resisted market relationships. He clearly believed that the grass economy, in its casual, hill-and-dale manifestation, threw nothing in the path of yeoman settlement but offered a tool in its service. As for the gentleman, Lorain advised that he give up the big stone house. A family will spend most of its time in the kitchen, anyway. The labor necessary to keep up a mansion (and the cost to keep it heated) is best saved by letting most of it sit empty. Under such an economy the poor could build their herds quickly without clearing more ground and at no cost; the rich could practice frugality that would make the farm a paying venture. The differences between East and West might mellow and so too the fire of emigration. Nature and Reason would be harmonized in the practices of husbandry.

Lorain hardly mentioned prices or returns from sale, even though those had become increasingly important to farmers within range of markets. One product in particular represented the desire for "get ahead" over and above the humble profits that protected permanence. Though rural people in certain places resisted the encroachment of the market into their lives, others relished that possibility and even fought for outlets. This is the story of the earliest union between agriculture and manufacturing, a subject that is also part of the story of improvement.

FLEECE AND BOUNTIES

Farmers investing in land wanted interest. Rich harvests represented one kind of return; fat and valuable animals represented another. In the first decade of the century, northern farmers hit upon a device that some hoped and others feared would transform the work of field and pasture into a profitable business with direct ties to manufacturing. Two prominent men, both with diplomatic connections, imported a breed of Spanish sheep in 1803, anticipating that the animal would redefine the wealth potential of agriculture. Rather than simply enhance the fertility of soils, merino sheep intensified the value of soils and in this way introduced a conception of improvement distinct from (though related to) manure and moral progress. The wool business redirected soil nutrients away from subsistence and local markets and toward manufacturing and more distant consumption. The gulf separating the supply of wool from the demand for it created a volatile economy among farmers, who sometimes spent stupidly for the animals and who stood to lose their fortunes if this four-legged speculation failed to play out. Never had a domesticated animal caused so much trouble, signified so much change, and freighted so much meaning about money and independence. In a brief time, a smallish breed known for fiber as fine as human hair linked together land and factories, altered the use and value of farms in the North, and became the indirect cause of the most significant sectional crises before the 1860s.

It began with two emissaries from the United States who saw the well-coated creatures as the genetic stock of a strong domestic economy. Even in the way they arrived in North America, merino sheep are the perfect example of the shifting meaning of economic independence tied to the use of land. Robert R. Livingston, member of the first Continental Congress and part of the committee that drafted the Declaration of Independence, took leave of his duties as minister to France in the Jefferson administration to investigate a woolly breed. Livingston had been in Paris since 1801 to prevent the retrocession of the Louisiana Territory to France and to negotiate for the sale of West Florida to the United States. Yet he had other things on his mind. He was thinking past the acquisition of territory to its sharpest use. A colony might be rich in resources, he believed, but a nation holds economic power through what it builds and manufactures. Spain had long considered the sheep too valuable to share with other nations and made their export illegal, but a small number had arrived in France. With an ambition to introduce improvements that "may add to the wealth of individuals, and, by forming the basis of manufactures, to the independence of our country," Livingston acquired two pair and had them shipped to New York.

In Robert Livingston's calculating mind the merino represented the highest aspirations of civilization. The sheep's characteristics—"overloaded with wool, slow in its movements, and possessed of no means of defense"—proved its centuries-long domestication and its evolutionary distance from wild beasts. Tiny hooves carried civilization forward through stages to its final form: "This little animal then, in losing its own wild nature, has not only converted the savage into the man, but has led him from one state of civilization to another; the fierce hunter it has changed into the mild shepherd, and the untutored shepherd into the more polished manufacturer."[68] Strange that manufacturing stood for the climax of civilization; stranger that sheep should be the catalyst. More specifically, sheep stimulated national maturation from subservient colony to Atlantic power. By some unexplained influence, sheep made people more sedentary and multiplied their wants for comfort, status, and so on. More wants meant greater dependence on those who satisfied wants, and this, asserted Livingston, was the condition of "polished societies." It was also the condition of complicated industrial societies.

Livingston returned to New York in 1805 to find his precious flock breeding with the common sort, having been neglected during his absence. No one wanted them, anyway—"novelties," said Livingston, need to wait "till the mind of the husbandman is prepared for their reception."[69] A flock imported by David Humphreys of Connecticut, also the booty of diplomatic business in Paris, at first found no buyers for lack of interest. Humphreys had fought at Long Island under Washington and had served as a spy. He learned the working of mills during a stay in England before the war and set himself up as a Boston merchant. When Humphreys rededicated himself as a Federalist after the election of 1800, Thomas Jefferson relieved him of the diplomatic post he had held under the previous administration. So he packed up his sheep and went home to Derby in 1802. There he founded the Humphreyville Manufacturing Company, which both raised sheep and milled their wool. In 1810, the venture earned Humphreys $500,000, and both he and Livingston reported selling full-bred rams and ewes for more than $1,000 each. The wool went for up to $2 a pound at market.[70]

The year 1810 marked the beginning of a merino mania in the United States. With visions of the fortunes they would make as domestic suppliers of the finest wool, American sheep keepers fueled a period of wild demand, while the supply of sheep remained strictly limited by politics and chance. Charles Oneill, an American merchant, found himself at the right place at the right time to come away with the hottest commodity in the Atlantic world. In August 1810 the junta of Badajoz, probably an ad hoc local government come to power during the war between Spain and France, confiscated 1,065 sheep from a flock of over 6,000 as war loot from the Count Montarco after he took sides with Napoleon. (Badajoz is in Spain's arid Extremadura—much like Arizona—on the southwestern border with Portugal where native sheep made a seasonal migration through the mountains.[71]) Oneill arrived in Badajoz and did his business "with very great difficulty," working through a third party named Guerra, "not wishing to be seen figuring in it." Days or hours after the merchant and his proxy reached an agreement with the junta, "no less than sixteen to twenty gentlemen appeared at Badajoz in pursuit thereof, provided with recommendations to the Junta from Marquis La Romana, &c. in expectation to be preferred; such as merchants in the British and American lines, and even Lord Cochrane, who came from England in a cutter for no other purpose than to secure this same flock, and send it to England. They arrived too late." By then Oneill had migrated with his quadrupeds across the border into Portugal and out of the way. Only 650 survived the ocean voyage from Lisbon to New York, where Oneill quickly put them up for sale.

Then came the War of 1812. Shut off from British textiles by embargo, Americans made their own cloth, and the new demand forced up the price of domestic cotton and wool. Fluffy sheep emerged from the war a prestige chattel, a symbol of American rebirth equal in significance, if not importance, to the Battle of New Orleans. A writer for *Niles' Weekly Register* called them "as imperishably beneficial as the renown of our military and naval heroes." In an article printed next to a story anticipating General Jackson's impending fight, another writer made this calculation: "10,000,000 Merinos—say they produce 3 lbs of wool each, and their annual value will be 60,000,000 dollars!—This is speculation, but gentle reader, not wilder than if a man, ten years ago, should have said that we might have our present quantity now. . . . Nay, this is moderate, compared with what has happened." Between 1803 and 1815 merinos climbed in popular opinion until they signified the flourishing of a free economy, the possibilities of individual wealth, the aspirations of American nationhood, and the renewal of agriculture in the old states. David Humphreys roused farmers in New Haven by telling them that whatever they touched they would turn to gold: "Away then with your Midases! . . . There is a mine in your land. It is so near the surface as to be reached by the shares of your ploughs. Explore it. Bring the treasure to light." A people stricken by turns with euphoria and anxiety named the merino "the National Animal."[72]

Merinos looked meek but grazed tough. Spanish sheep brought new vigor to the old states because they could be carried profitably on the most ragged, tired ground. They had had a long history as free-range foragers by the time they reached North America. Their ancestors inhabited the lowlands of the Iberian Peninsula in Roman times. Shepherds noticed that they thrived during seasonal migrations—wintering in the south, moving up-country for green shoots as the plains dried out in summer, then returning with the onset of snow. Sheep walks measured about seventy-five meters wide and went for hundreds of miles—a kind of interstate pasture. One route passed by the city of Badajoz. Americans used merinos to make the most of threadbare acres and marginal spaces. This animal, claimed the Merino Society of the Middle States, would "hasten the revival of agriculture," a process "by which the worn-out farm will be speedily renovated, the flocks will be increased, their immediate reward will be sensibly felt, and promptly acknowledged by the now famished fields; yes, fields which but a short time ago were a continued drawback to the farmer, may under judicious introduction of proper flocks, become permanent sources of wealth to the owner." Some looked to "the mountainous and hilly districts of our state, at present of comparative little value," to become the "most profitable sheep grounds." Sheep improved barren ground in the first and oldest sense of the word, by bringing into better account part of the country "which otherwise could not generally be cultivated at all." On scrub and stubble, "where no other animal would be maintained with equal profit," they not only thrived but multiplied. Another observer even claimed that old improved land well set with bluegrass "is much better for rearing sheep than the richest new lands."[73]

Not everyone shared the conception of merinos as denizens of outlying fields. The British merchants and landlords who schemed for them had no intention of turning them out to the wastes (or of flocking them to high meadows), and some believed that the more money spent on fine-fleeced sheep, the more wool they would grow. Sheep keeping could also be an intensive land use, and here again the British exerted great influence. The New York Board of Agriculture noticed vast differences in the ways American and British farmers nourished this livestock. "Everybody knows that sheep live upon next to nothing; . . . sheep are owned as a mere convenience, supposed to cost little or nothing to maintain, and are placed at the very fag-end of the concerns of the farm." But English farmers brought them inside the hedge; that is, they extended the boundaries of cultivated spaces to encompass them. Noted the New Yorkers with regard to the typical British tenant, "Sheep indeed, are the basis upon which all his operations are raised. The whole of the admirable system of convertible husbandry . . . are but so many steps preparatory to carrying on his sheep to the greatest possible state of perfection."[74] Some Americans fed the animals as they did cattle and collected their dung. Some obeyed the most exacting rules of high keep: never put them on cold soils or wet bottoms; never leave them unsheltered in the winter; and weed their pasture of species that make them sick.[75] Added up, merinos coddled to this extent commanded more labor than they were worth—a highly leveraged investment, a downright speculation.

Regardless of how wool growers kept them, sheep became a force in the landscape. Their importance derived not only from the lines they occupied in account books but from the niches they occupied in the ecology of farms from Vermont to Virginia to Ohio. Calculated one county convention of Pennsylvania farmers, "Our cleared land is estimated at 250,000 acres, capable of maintaining on an average, two sheep to the acre, without rendering our population dependent on others for those agricultural products which we consume, and now produce within ourselves." In 1827 these farmers were thinking about land use almost exclusively in terms of sheep. "According to this estimate, we can keep 500,000 sheep, yielding one million five hundred thousand pounds of washed wool, which will leave, after deducting the quantity necessarily consumed by a population of 50,000, a surplus for sale, of more than a million pounds."[76] By 1837, within ten years of these estimates, sheep had taken over entire counties in the North. New York farmers grazed an amazing 4,261,770, with two counties combined claiming over 400,000—more than the total for the state of Massachusetts (373,322). Virginia stocked 160,000 in only two counties, with an estimated 1 million for the state as a whole.[77] The most significant numbers may be the smallest: the density of sheep in Vermont never seems to have exceeded between 1.5 and 2 per acre. In other words, the state's 1.7 million sheep (compared with only 384,000 neat cattle) represented an enormous dedication of space. By the late 1820s they had become an unmistakable geopolitical force in the northern countryside, and Pennsylvania farmers sensed the risk they had taken: "More than one-fourth of our cultivated surface is occupied directly or indirectly with sheep. What would be the effect, if by the destruction of our flocks, this land was again thrown into the cultivation of grain?"[78]

No one wanted the answer to that question. In just a brief time, the merino mania created the illusion of a tight equilibrium among three forces—manufacturing, general husbandry, and wool raising—that soon tipped over. When trade resumed after the end of hostilities with Great Britain, the price of fine wool dipped in 1816, again in 1820, and then began a slide to its up-and-down low point between 1826 and 1830. Broadcloth declined 40 percent, from $4 a yard in 1825 to $2.25 a yard in 1829. Stories of the bust appeared for the next fifteen years. A visitor to the countryside near Baltimore called on a friend who insisted that he stay the night. The city man enjoyed a meal of ham and chicken, but his host made apologies for the food and said that if he would stay another night they would eat a freshly killed merino sheep. "No better to him than others" after the price decline, wrote the visitor. The farmer had paid "from one to three hundred dollars a piece for the breed, expecting to make a fortune by them; but like many other projects, it ended in smoke." Expectations of wool at $1.50 a pound hit bottom at 50 cents. Why not raise wool for your family and never mind the market? Too much trouble, "and when done, [it] did not look so well as imported cloth." "When I want a coat," said the farmer, "I just go to the store . . . and pay for it."[79] So much for independence.

Looking back thirty years on his brief employment as a store clerk, the writer and publisher Samuel Goodrich remembered the merino craze. His employer, shiftless and unfit for business, "sought to mend his fortune by a speculation in Merino sheep—then the rage of the day. A ram sold for a thousand dollars and ewe for a hundred. . . . Fortunes were made and lost in a day, during this mania. With my master it was a great cry and little wool; for after buying a flock and driving it to Vermont, where he spent three months, he came back pretty well shorn—that is, three thousand dollars out of pocket!" A New Yorker writing for the Board of Agriculture in 1821 recorded that the price for merinos finally equaled "the value of the commonest sheep of the country; nor have I heard for some years, of a merino sheep being sold for any price, unless fat and fit for the butcher."[80]

The rise and fall of the merino suggested big changes in the American economy and in the commercialization of agriculture. The merino was, first of all, the first get-rich-quick craze in the countryside, the first product farmers bought in an attempt to read the market and make an easier buck. Long before mineral rushes enticed farmers to go prospecting, fancy sheep planted the idea that "profits are to be gained by lucky hits in the lottery of chances; in preference to the slow but sure rewards of . . . prudent management." Even after a plummet in their value that pulled rural families down with them, merinos continued to attract defenders who did not flinch at the notion that agriculture would always be tied to manufacturing and distant markets.[81] An astonishing trend emerged from the nexus of wool growers and the water-powered mills that fabricated broadcloth: a unity of interests. Here was a rare case. In certain districts of New England, New York, and Pennsylvania—all of a sudden—farmers and the makers of textiles had ties, not simply because they sometimes shared a labor force but because they lived on either end of the same agricultural economy.

Factories purchased a raw material that came from land in farms. Anything that diminished demand for the finished product also affected farmland in a chain reaction new and shocking to the people of the 1820s. As the founders of the Merino Society of the Middle States put it, whether raisers or manufacturers of wool, "our mutual interests [are] happily combined." The constitution of the Merino Society of the Middle States virtually cast the enterprise of wool growing as an integrated facet of a larger industry and stretched the old meaning of *independence*. "Improvement in manufactures, cannot long exist, when, what are called the raw materials are not generally raised. . . . Without the raw material, manufactures, and commerce, we should possess nothing more than the name of Independence."[82] New

York wool growers called them "inseparably blended, and dependent on one another," and claimed, with little experience to counsel them, that "the extensive establishment of manufactures creates the best and most stable demand for all the products of the soil. Agriculture and manufactures, therefore, must flourish, or decline together."

There is no sense of ambivalence in these statements, but rather the idea that farm and factory proceeded as equals. Actually, the one gave birth to the other, and rural mechanics performed the delivery. Far from being stragglers on the road to modernity, farmers invented some of the first mass-production machines. Born into a rural family, Oliver Evans apprenticed to a wheelwright in Newport, Delaware, before inventing the automated gristmill in 1784 at the age of twenty-nine. The machine that Evans built virtually eliminated the human labor needed to turn wheat into flour. From that moment on, milling cost less wherever water and capital came together. Technological change this penetrating never pointed in one direction. Larger and more complicated machines required unprecedented capital investments and more sophisticated management of looms, materials, and people. The new process began to distinguish the mill from its rural origins as a machine scale of work and energy emerged from a human scale.

Early on, though, wool growers and woolens manufacturers could be the same people. With the Atlantic trade halted during the War of 1812, water-powered mills appeared in the countryside throughout the Northeast. As early as 1820 Pennsylvania counted 123 establishments and more than sixty thousand people engaged in manufacturing, with some counties nearly split between their spindle and plow populations. The county with the most people engaged in husbandry (Montgomery, with seven thousand) still had twenty-five hundred in manufacturing. In the eastern part of the state generally, the number of people in factory work equaled 50 percent of the number in agriculture.[83] Though New York and New England also saw the number of people so engaged take off during and after the war, the quality of the mill on the Brandywine and the Delaware differed from that on the Merrimack. Rather than giant textile factories operated by armies of dislocated rural women (as in Lowell, Massachusetts), industry in

Pennsylvania tended to be owned and operated by merchant millers and located close to fields. Free white workers moved from flour milling to farming and back again, depending on their best opportunity and the season. The rural origins, small scale, and petit ownership of both flour and wool milling lowered the volume on "the coming of industrial order" to the American countryside. Agriculture and industry in Pennsylvania relied on the same products and the same labor.[84]

The sheep keepers of southeastern Pennsylvania saw no contradiction in a republican nation built on agriculture and manufacturing, but their participation in the new economy exposed them to speculative risk.[85] Calls for relief from both sides of the thread moved faster than a steamboat to the statehouse, where a committee voted 22 to 9 in favor of lobbying Congress for a Woolens Bill in 1827.[86] That bill failed, but a convention of delegates from thirteen states, including three from the South, met at Harrisburg later the same year and signed a memorandum to Congress on August 3, in which they wrote: "Forty millions of manufacturing capital, together with forty millions of farming capital, composing this great national concern, for want of adequate protection, have lost half their value. It is in the power of congress to relieve it from present distress and jeopardy, to prevent its utter ruin which is imminent." Disclaiming any sectional or individual interest, the convention penned a bill for protection that, unlike the first, covered raw wool as well as goods composed of wool. It served as the blueprint for the Tariff of 1828. The Harrisburg Convention, as it came to be known, exemplified the liberal republicanism at the core of the emerging manufacturing economy. Names affixed to the memorandum included Mathew Carey of Pennsylvania, Jesse Buel of New York (a vice president of the convention), and Hezekiah Niles of *Niles' Register*.[87]

Always takers and not makers of prices, farmers usually felt their lumps in a downturn and evened out a bad balance by taking land out of production. In the case of wool, however, the source of market trouble came from abroad and could be offset by government through one of the oldest forms of commercial law: tariffs. Burden the importation of English wool and woolens by attaching a duty at the wharf and the domestic price would go up. The United States could birth an industry and protect the value of sheep and farmland in one stroke. Farmers and manufacturers formed coalitions and buckled down to lobby for the passage of a bill. With the destruction of its markets, they said, "husbandry languishes." Wool growers and their allies in industry advanced the argument that "a large amount in cash and labour has been expended in introducing, increasing and improving our flocks. This amount must be lost to the enterprising wool growers and to the community, unless further protection is afforded, for it is not to be expected that the most patriotic will continue the business at a certain yearly loss."[88] Responding to anti-tariff Philadelphia merchants known by their favorite hangout, a Chester County farmer wrote, "We believe a Breakwater would be a very good thing to protect ships; and we can't for our lives see why a Break-woolens wouldn't be a good thing, to protect our sheep. They certainly haven't shown any reason, at the Coffee House, why an honest, industrious farmer shouldn't have a protection for his sheep."[89] Never before had farmers supported an industrial policy.

The members of the Harrisburg Convention had no trouble grasping the ground-level benefits of the policy they sought. Here is an example of land-and-tariff reasoning. A factory with 160 workers processed 100,000 pounds of wool each year, an amount representing the product of 35,000 sheep grazing on 115 farms, each with 200 acres, 300 sheep, and a family of six. Represented by about fifty large factories (a fair estimate of the number operating in the United States), the industry as a whole would have "employed" 5,750 farm families and a rural population of 34,500 on 1,150,000 acres. Should the woolen chain ever break and farmers turn all these acres back to plowland, the changeover would ripple through the grain market and deepen the depression. Asserted another convention: "The agricultural [interest] is so closely connected with the wool growing interests that it must seriously feel any depression which occurs to the latter. . . . The wool grower is now a customer to the grain-grower, instead of his rival—a consumer instead of a producer."[90] Within the rural economy of Pennsylvania there emerged the rudiments of industrialism, defined by a profound interdependence between agrarian and factory production. Merinos changed the countryside in ways that Robert Livingston could not have imagined. They altered patterns of land use, concentrated capital in mills, and gave northern farmers different interests from southern planters.

For northerners in sheep counties, the tariff functioned as a form of political improvement—a bill for protecting a particular rural land use and an industry increasingly favorable to the national balance of trade. A New Yorker writing at the time said as much, recasting the role of the agricultural society toward "providing prompt and ready markets" and calling upon "the powerful shield of government" to aid united farmers and manufacturers. But sheep languished in the South. A touring Frenchman related a noonday interview he held with an English farmer recently hired to manage a Virginia plantation: "Many sheep cannot be kept in summer. Little mutton or wool is wanted, and were they generally marketable, there is no winter food for sheep." Wool had fewer uses in a warm climate, and planters did not like to devote time or acreage to livestock. There is no way to overstate the gnashing of teeth, the fury, that duties provoked among planters with different economic interests from northern farmers and manufacturers. When the tariff bill of 1828 passed both houses, it imposed the highest duties ever passed by Congress—4 cents a pound on raw wool and an additional 40 percent of the market value of woolens, increasing to 45 and then 50 percent over the next two years. Planters in local associations sent letters to Congress denouncing the bill in language they might have lifted from *Arator*, calling it "the destruction of agriculture, for the patronage of Manufacturers."[91]

A great political battle widened in which Vice President John C. Calhoun rendered himself unelectable to the presidency by propounding the "Doctrine of Nullification" as a brief on the Constitution. He and the radical republicans declared this "Tariff of Abominations" a dead letter within the borders of South Carolina and precipitated a standoff that lasted from 1828 to 1832, almost provoking Andrew Jackson to dispatch the Army to enforce federal law. It was the most serious sectional confrontation and the purest constitutional crisis before South Carolina seceded from the Union in 1860.

Not just two diverging sections but two agro-environments and two settlement regimes met in the controversy over the Tariff of 1828. While this narrative has dwelled on how Pennsylvania abounded with manure and premium animals, it has yet to consider how South Carolina did not. It is no accident that the two states fell so neatly on either side of the tariff issue. Pennsylvania: portly and snug, seemingly incapable of producing a leader with guts and stature (including James Buchanan). South Carolina: scrappy and fierce, made up of dusty planters bent on reinstating the Articles of Confederation. The two states represented dire opposites in climate, cultivation, labor, and economic diversification. Leading planters repeatedly blamed tariffs for the economic distress they suffered, when soil erosion and competition from the booming states of the Southwest presented more likely causes.[92] The drive for improvement played out differently in South Carolina, where the ligaments connecting soil and society strained under plantation land use.

OLDFIELD

South Carolina stood alone against the Union, a mistake its ruling planters would never make again. Feelings of loss and fury had more than one cause, and the tyranny of an overreaching government was not the only source of dread in 1832. South Carolina's ecological degradation had been apparent for decades, but it was the emigration of the 1830s and 1840s—some of it a direct consequence of the tariff and the Nullification Crisis—that finally jarred the big-shot planters like an earthquake offshore. If the chain linking husbandry to society fell apart anywhere in the United States before the Dust Bowl on the Great Plains, it came closest in South Carolina after 1832—a period of enormous stress that tested the assumptions behind improvement.[93]

Signs of an ailing tenure in the countryside stunned one observer with uncommon eyes for landscape. Edmund Ruffin followed the

Cooper River out of Charleston during his first week of an agricultural survey of South Carolina in January 1843. No one prepared him for the vicinity: exposed subsoil, second-growth loblolly pines "of mean size and appearance," naked land, "and the houses all show decay and dilapidation." Ruffin could discern no attempt "to improve or even cultivate the land." The sickliness of the country drove some of the inhabitants to higher ground during the wet season, but they left little to return to. Many planters lived rich and well on land that produced short-staple cotton, yet Ruffin's diary is punctuated with references to poverty and abandonment. Pausing along the Ashley River in early February, he came upon plantations like unburied bodies, each a "melancholy scene of abandonment, desolation & ruin." The crops failed to pay for the work and cost of cultivation—a little rice, some cotton, a patch of corn—but the inhabitants purchased hay from up North and made most of their dreary income cutting wood and selling it in Charleston. "There seems to be no thought of doing better," Ruffin said of them.[94] On James Island he confused soil and water for a moment while watching dry tilth taken off by an updraft; the grains lay down across the roadbed "in ripples as the waves on the sandy beach of a wide river."[95]

Plantation land use leaned the white residents of South Carolina toward abandonment. This is a gross generalization, since many planters succeeded in districts all over the state. The planters considered here at least succeeded in staying put. But page through the *Southern Agriculturist*, or *Soil of the South*, or the *Farmers' Register*, and the grinding tendencies of corn and cotton become impossible to ignore. Decades of writing in the rural press and in the private papers of improving planters reveal a tangled negotiation between a desire for soil restoration through fodder crops and manure on the one hand and the labor and land economics of planting on the other. There is nothing more fraught with subtleties and contradictions than the antebellum plantation, and any conclusive attempt to understand it would require more detail and greater space than I provide.[96] The following is not a history of South Carolina during the spread of cotton culture, nor is it an account of either the lowlands or the Piedmont. It is an object lesson in the struggle for permanence. Thomas Cassells

Law of the Darlington District (in the middle country) and David Golightly Harris of Spartanburg (high in the Piedmont) worked for decades without uprooting their households and slaves for a new country. Throughout the 1830s and 1840s an out-migration on a scale never before seen in the United States raised the strange prospect that settled society in South Carolina might fail its mission, yet the practices of these planters (the work of their slaves) tell the story of a brutal agricultural regime, a burdensome environment to till, and a hard-won permanence.

The lines connecting conservation and improvement are more murky in the South. Northern improvers were heir to the prickly high-mindedness of Puritan culture. Planters joined the same kinds of associations and wrote and urged the very same practices as northerners (with important modifications, as we'll see shortly), but they urged differently. Southerners worried most about their declining plantation independence as a foil to northern tariffs and antislavery. Edmund Ruffin best resembles a southern George Perkins Marsh, but where Marsh eventually extended his concern from the denuded hillsides of Vermont to the forests and waters of the inhabited earth, Ruffin remained a southern patriot to the bitter end. Not only that, but plantation land use contained elements dangerous to restoration, though Ruffin refused to admit it. The grass and cattle husbandry could not be land extensive; rather, it dedicated the most precious space on any farm to fertility. Cotton planters rarely attempted it for this reason, but they did attempt a regional compromise. Ruffin and the few planters who made motions toward a restorative system before the Civil War wanted permanence of home and landscape just as northern improvers did, though they never seemed to have recognized how slavery and cotton combined to frustrate restoration at almost every turn.

Thomas Cassells Law and Mary Westerfield Hart assembled fifty-five hundred acres in Darlington, in the Sand Hills just below the Fall Line that distinguishes the Piedmont, near the border with North

Carolina. It appears that the portion in active cultivation at any given time in the 1840s consisted of about four hundred acres worked by about thirty slaves.[97] The formation of the family's homeplace followed the inland pattern. The settlement generation started out in Great Britain, paused in Pennsylvania and Virginia, then crossed North Carolina to reside in the Sand Hills during the 1790s. The same could be said of Patrick Calhoun and his family, who arrived in the 1750s in the Piedmont Abbeville District, where John Caldwell Calhoun was born in 1782. Thomas Law was also born in South Carolina, in 1811, a decade after his parents arrived. Law's neighborhood of Society Hill looked like backwoods even into the 1820s, when the engineer and surveyor Robert Mills described houses "built without any regular plan in the woods . . . and so scattered, that, as you ramble, you come upon them unawares." The dwellings lacked neatness, ornament, and a coat of paint even thirty years after their inhabitants had begun to make money from cotton. Land in Darlington displayed the common disparities of the country—good tilling soil in the river bottoms and associated lowland swamps of the Pee Dee River interspersed with pine barrens. Offering words for improvement, Mills suggested "a change from the planting to the agricultural system; or, in other words, in place of impoverishing to nourish the soil."[98]

Planting and farming were not the same, and attempting both at the same time was like being pulled apart while holding the reins of two wild horses. As a diversified farm, Law's place was a feeble reflection of James Pemberton Morris's in Bucks County, Pennsylvania: cotton, corn, peas and potatoes (in what may have been a household garden), a little wheat (twenty-seven acres), even less rye (four and a half acres), and even less oats (just half an acre). That was all. Law mentioned hogs in his plantation diary, but never in four years did he write specifically of cattle. The words "fodder crops" appear, probably a reference to stored hay from the wheat or rotting cornstalks along with assorted litter, like what his slaves raked up in the spring of 1844 from fields and creeksides. He kept no root crops, nor did he give a hint of seeded meadow, yet every spring Law hauled and spread manure, almost certainly a dung produced from some mixture of wild pasture and whatever else was available. He did not manure every field every year but followed a long cycle. In March 1844 he directed slaves to rake it on ground he called the Williamson field for cotton and also on the Large field. They hauled it for potatoes in the garden and for corn in the Leslie field. He noticed crabgrass growing on the "manure heap" and also referred to "lot manure," suggesting the presence of an uncovered pen where animals made dung. Through all his troubles and annoyances, Law remained in Darlington his entire life, first because he wanted to and second because what he did to build the resources of his soil worked well enough to counteract the worst tendencies of his cultivation. Without a doubt, he would have endorsed the Newberry Agricultural Society when its members rejected the idea that "cotton culture is incompatible with the improvement of the soil."[99]

Yet although Law represented a tenacious minority of land-conserving planters, the manure his slaves spread every spring did not maintain the number of acres he planted. The proof punctuates his lengthy journal. About every four years Law's captives cleared a tract of forest and planted corn after burning the ground. This "new ground" followed a regressive series of rotations over five years between 1843 and 1847 without rest or manure—corn, corn, corn, cotton, cotton. After the last crop Law recorded "burning logs and picking up trash" on a tract he named "Big new land," where the same pattern continued. Then the next year, 1848, he began to cultivate a tract called "fresh land," distinct from the other two, where he began a cropping cycle with cotton. The pattern is unmistakable: Law needed to cut the woods in order for his plantation to function. Rains flooded the lower fields in 1847, drowning corn and cotton, and may have sent him up a forested hillside, where he found humus for greater water retention and dependable fertility at a time when he sorely needed both. But a far more likely reason for the clearing lay in Law's inability to make manure of the quality and in the quantity that would have restored his worn soil. Slavery too was directly implicated in the extensive and destructive practices of cotton. Under certain circumstances both factors tended toward exhaustion and emigration.

Planters knew all about convertible husbandry and never doubted its results. They simply refused to pay for it. Nothing could be more important than manure, wrote James Henry Hammond for the Barnwell Agricultural Society in November 1840, the only problem was its cost. Hammond's advice was to keep it cheap: "A great deal may be made on every plantation, without much trouble or expense, by keeping the stables and stable-yard, hog and cow pens, well supplied with leaves and straw." Keep corncobs too, throw them in with the "sweepings from negro and fowl-house yard[s], and the rank weeds that spring up about them, collected together and left to rot." But Hammond stopped right there. To walk one more step in the same direction by forming "a regular force detached to make manure, at all seasons, and entirely left out from the crop," was a serious decision for a planter to make, and Hammond did not recommend it. "These operations partake more of the farming and gardening, than planting character . . . if anything like an average of past prices, can be maintained, it is certain that more can be made by planting largely, than by making manure as a crop."[100] This from the famous land-thinking governor, friend of rural reform. In fact, Hammond emerges as a common planter, spending fertility while cotton was high no matter the long-term consequences.

Another writer said much the same, pointing out the prohibitive cost per acre of first-class dung: "No one will, for a moment, doubt that stable or compost manure, is to be preferred to any other kind; but when we contemplate the great quantity necessary per acre, we are driven to one or two alternatives, either to abandon the process of manuring, or search for a more convenient plan." What planters did not want to do, nature, conveniently, had done for them, or so the writer happily informed his readers—that there lay, "at hand, an excellent substitute," better known as cornstalks, fallen leaves, cut straw, and the mast raked up from the forest floor.[101] One from Mississippi shot down the whole notion: "Our want is the cheapest system of enriching and preserving large plantations—for it is outrageous humbuggery to talk of hauling manure over them." Improvement the way John Lorain and James Pemberton Morris understood it came to the South at least twice cursed. First, it asked planters to reallocate land and available capital away from the commercial staple crop and toward restoration. Spacious barns, sown meadows, and complicated rotations all required investment for gains less immediate than this season's bolls. Second, and even more limiting, it required the diversion of slave labor from cotton. Some planters found ways around both problems, and some rice planters developed sophisticated systems to manage a great variety of crops and landscapes, but the majority made no attempt.[102] Fresh ground indicated a failing tension between wearing out and putting back, a vision of exhaustion in slow motion.

Scarce animals also told of a poor economy. Observers on the ground in the South recorded the lack of domesticated animals—in striking contrast to the abounding northern herd. John Taylor, never one to remain silent about the failings of his fellow planters, asserted that a lack of cultivated grasses translated into a lack of animals and all that animals provided: "A thin soil, exposed to hot and dry summers, not only prevents our lands from clothing and nourishing themselves, but has at length rendered them even unable to raise working animals for their own cultivation. . . . A remedy for this state of things is necessary to stop the emigration from Virginia, and to prevent its ultimate depopulation." This shameful laxity, this utter neglect, moaned a planter from the environs of Greenville, "may be seen in almost every horse, cow and hog we find on the farms in this District." In a letter to the New York State Agricultural Society, a correspondent saw few sheep and said that horses, mules, and all the good cattle came from Kentucky: "With very few exceptions, the woodlands furnish the only pasturage, and the cattle for the most part are diminutive and unproductive." Without more attention to feed it would be a waste to import better cattle, he said, and the only way to make more feed was by "farming," a system that "requires one-half of the farm to be cultivated in some of the cereal grains, while the other half is under pulse [legumes], roots, cultivated herbage, or simple fallow."

The paucity of quality stock may be part of a larger chain of causation related to the environment. The great artificial grasses perished in the highly acid soils so common where cotton flourished. Planters unaware that lime would sweeten their land, or at pains to dig it out where it appeared naturally (in the form of fossilized seashells, called marl), refused even to attempt grasses. It is not true, however, that no forage or meadow grasses took hold in South Carolina. As Governor John Drayton documented in 1802, a few planters near Charleston practiced mowing to take advantage of the value of fresh-cut hay, though he suggests that they mowed the native crabgrass, "the grass which is most attended to in Carolina." A Colonel Hill of York planted twenty acres in "good grasses" that included red clover. Still, the major European forage crops—which formed the biological foundation of convertible husbandry—did not do well in the lower South without attention to acidity. Rain at mowing and curing time only made a bad situation worse. As the historian Jules Rubin has argued, the various negative effects of climate prevented "a productive combination of crop and livestock farming which had mutually beneficial effects," and deterministic explanations for the distinctiveness or the backwardness (or the distinctive backwardness) of the South have been offered for a century.[103]

Yet the American South was not equatorial Africa. Keep in mind that the "package" of domesticated plants and animals imported from Eurasia—including wheat, sheep, oats, and cattle—became established in the South. Compelling as they are, environmental explanations cannot account for a political and economic regime that depended on the waste of land. William Tatham, an English resident who paid close attention to agriculture, voiced his amazement that a country so well adapted to animal husbandry (Virginia) neglected its cattle so badly: "Nothing could be more easy, in a country clothed with timber, . . . than to house them comfortably; nothing more interesting, than to feed and improve the breed where nature has furnished most of the necessary food; nothing easier, than to practice *management* where every farm has sufficient space . . . all these points are however neglected, because Nature has been so profusely bountiful in bestowing the *mediums of makeshift*!" Planters overwhelmingly believed that no natural impediment prevented their state from attaining prosperity. Said one in 1839, "Let no planters . . . despair of making at least 1250 lbs. of clean cotton to the hand. It can be done on almost any land in the State."[104] Robert Mills scolded planters for their addiction to the one crop that drained the resources and human energy from every other: "There is not a finer grazing country in the world than South Carolina; and were attention paid to the raising of cattle, sheep, goats, hogs, horses, mules, &c., this state might supply itself as well as all the West India islands . . . ; but every other object gives place to cotton."[105]

A dead-leaf diet did not deliver restoration. It might have fattened animals when offered to them in large enough quantities, but in the long run it slowed their growth because it lacked the proteins needed to build muscle. Dung from forest debris possessed little of the power of dung from rich grasses like timothy, so even those planters who manured did themselves little good. Even worse for planters hoping to get by with what the floor and riverside gave them, cattle at large inflicted astonishing injury to land. When they were allowed to erase the boundary between feral and cultivated spaces, cattle carried every weedy species they came across back to the plantation and injected it into the soil through their dung. Finally, a limited supply of field stubble and mast (acorns and nuts) limited the size of herds and thus the total amount of manure that could be produced.[106] If the nineteenth-century farmer was just getting by with half a cow per acre, then Thomas Law needed from 150 to 200 head to keep his cotton and corn in top production. It is extremely unlikely that he kept a herd of that size; he certainly did not feed that many out of his own resources. No wonder he cut his woods. It was all he could do to keep from falling behind in production, but he still fell behind the needs of his soil. Law held off emigration throughout the 1840s and into the next decade by a combination of dung and clearing on the enormous tract of forested land he and his wife inherited and assembled over their lifetimes. He conducted a slow shift through brush and timber, financing expansion with fertility he did not have to pay for, like emigrating while staying in place.[107]

It is tempting to view the plantation through northern eyes, but that would be a mistake. Northern improvers, wearing lenses they borrowed from English farmers like William Strickland, saw every abandoned fallow as a weedy waste. But what's so wrong with emigrating while staying in place? What if Thomas Law really *did* rotate his four hundred planted acres through fifty-five hundred over a long period of time? Why call that regressive? A more informal system of cultivation—a slash and burn, or swidden, in which planters used fire to prepare a clearing, cropped for a few years, and then let the land return to bush—was entirely stable and in fact, to the extent that it did not require much capital or animal stock, was more secure financially. James Mercer Garnett of Virginia defended the three-crop rotation as the epitome of practical economy and asserted that it worked just fine: "In defense of our 3 shift method, I will state the following facts . . . from my own experience: The first is, that my fields, cultivated in 3 shifts, but grazed only by sheep, have obviously improved considerably in a few years." Under the bad old system of rotation and naked fallow, his fields produced sixty bushels of corn to the acre, and he brazenly recommended planting wheat on corn stubble without rest or dressing, conceding, "I am aware that I am now treading on very debatable and debated grounds; and I am not without apprehensions." And as for the dead-leaf diet, why should a planter have worried about that? Call it an economical way to add fertility at next to no cost, with little labor expended. All told, assuming enough land over which to shift a large cultivated space, with a ten- to twenty-year fallow for each field after about five years in tilth, it would be possible to disregard the restorative or "farming" system for a less intensive "plantation" method.[108]

This is why the South looked so different from the North and why visitors called it trashy. A traveler on foot through oldfield would have waded through broom grass knee-high while stepping over half-burned logs and stumps never removed, then halted and doubled back rather than enter the bushy bramble and weedy pines still too stunted and wiry to be called an understory. A closer look would have revealed species that thrived on disturbance, soon to form new forest stands different from and less diversified than the first. Oldfield was a cultural artifact, and to people unable or unwilling to read it, it simply looked ugly. Speaking of Alabama settlers in the third-person singular, the English traveler James Stirling said, "Even his cultivated patch is, in many places, disfigured with stumps, or what is still more unsightly, girdled trees, which rise like great naked, death-like poles, all over the surface of the ground."[109] Observers of Stirling's sort almost never wondered or inquired about the benefits of the practices they deplored. Planters graded hillsides in horizontal rows to form terraces; they set cowpeas rather than English clover to transfer atmospheric nitrogen to soil; they intercropped cowpeas with corn in a manner reflective of American Indian practices. Along with swidden and the traditional three-crop shift, these regional adaptations cost very little. Planters wielded them knowledgeably, even if their efforts fell short of full restoration. In other words, there could be a fine art to "falling behind." It allowed them to conserve labor time and capital by employing the regenerative power of the forest to re-create their land.[110] It would never be highly productive, and its diminished capacity always became a dreadful question whenever cotton prices fell, but as long as planters mustered the minimum and had sufficient acres, it would be stable.

In this light, the dunghill doctrines look severe and unforgiving of the feral tendencies of rural landscape. More ominously, they seem to mark the beginning of a monolithic conception of farmland management. This way of thinking, a pillar of agricultural science and government policy as they emerged in the next century, chipped away at local differences by regarding cultivation as a process that could be isolated from the greater environment. The geographer and historian Carville Earle further contends that the new husbandry did not offer long-run stability for the South. Constant cropping through fertilizers, deep plowing, and clean-tilled fields "initiated an unprecedented period of environmental destruction" in the region by accelerating erosion. In this view, proponents of the method become forebears of the chemically dependent, highly erosive habits that characterize cotton production in particular and industrial agriculture in general to this day.[111] It is an important view, but there are problems with it. The grass and cattle husbandry never comprised a single set of grasses, animals, and courses that could be promulgated but instead withstood wide regional variation. Improvement also included green manures consisting of crops like rye, clover, and cowpeas, which preserved topsoil and did not constitute "clean till." Still, here are some thorny questions to be revisited in Chapter Three: Did improvement draw the starting line for the technological scramble that led to industrialized agriculture? If it did, can grass and dung be cleared of all charges?

The positive interpretation of southern swidden, however, was not even popular in the South. Small plantations could not have accommodated a ten-year rotation of enough cleared land to pay the master. "Look at the exhausted fields that are laying waste," said a frustrated colonel at Columbia in 1818. Proprietors were obliged to cultivate such sterile ground "because they have not a sufficiency of woodland to justify them in clearing new and fertile ones."[112] Slash and burn requires an abundance of unbroken land as a sink, just what many planters lacked. Besides that, southern improvers almost never honored swidden and, like northerners, refused to see land in long fallow as anything other than a problem. James Hamilton Couper, the manager of a giant Georgia sugar plantation, laid out an intensive course of convertible husbandry that made no apologies to northern farming. Said Couper, the conscientious planter "will not allow his soil to clothe itself with a spontaneous growth: for then that fertility is expended in useless weeds which might be converted into valuable crops." Seeds leftover from weeds and second-growth forest threatened future crops.[113] Most dramatically, the members of the Agricultural Society of South Carolina worried that oldfield fallow represented the regression of civilization. No elegant compromise with nature, land in long fallow was a capitulation to forces better kept at bay: "Complete exhaustion will at length compel to a total abandonment. It is easy to foresee that an extensive practice of this kind will, in the course of a few years, greatly impoverish a country, and finally convert it almost into a wilderness."[114] Swidden may have functioned as a sustainable practice over a large and extensive landscape, but it failed where settlement became close.

As soon as seeds fell across turned up "new ground," problems of a different nature, having to do with the intersection of land and slave labor, took over. Cleared spaces were too large to care for, and according to the upside-down logic of cotton, they needed to be too large. "With us," wrote a tired practitioner, "where the estates are so large, the question of renovation is indeed one of difficulty. . . . This is the great and crying evil of our day, and under such a state of things man must be ruled by his estate, and is frequently a greater slave than he for whom money is paid."[115] Why would a planter feel enslaved by the size of his plantation? What forces drove the need for expansion and at what costs?

In June 1843 an intrusive grass attacked Thomas Law's cotton. A hearty crop of the fiber required perfectly clean rows, so the grass presented a serious obstacle and demanded a great deal of work. Their labor already spread thin in other tasks, his slaves could not be regrouped to fight the grass. Paralyzed by the emergency, Law realized his error and wrote his own rebuke:

> June has been . . . altogether the most difficult month I have ever known, to cultivate a crop successfully—the Grass has grown beyond any thing I ever witnessed, my whole cotton crop has been given with it from row, to row, never had half as much grass in my life—which is a general complaint—Farmers should not over plant themselves, for a small crop well cultivated will make as much as a large one badly tended and there is some pleasure in having a crop in good order. . . . 15 acres is enough to the hand and 30 to the horse—Cotton middles should never be left long without ploughing them out [,] never till June. Cotton generally looks well [,] some what backward and checked in growth by grass.[116]

Law blamed his inability to fight the infestation on a lack of labor, and his lack of labor on the far-too-many acres he planted that year. Most revealing is his statement that 15 acres is enough "to the hand," meaning that he estimated his planting and harvesting by the number of acres each slave could work, not by any estimation of the nutrients in his soil. Law's allocation of acres per slave was more than double what another source recommended as moderate. A correspondent in the *Southern Agriculturist*, writing on the "general management of a plantation," recommended 6 acres to the hand with 20 slaves, for a cotton field of 120 acres. If Law employed 20 of his 30 slaves to work cotton, each with 15 acres, then he had 300 acres of short staple.

With that much under till Law had no margins for error but plenty of opportunity for it. He took the planter's gamble by thinking with his labor and not with his land.[117]

To the hand. If the land and labor economy of the cotton South could be held in a phrase, it might be this one. Like northern farmers, cotton planters considered labor a limited quantity and land more or less limitless. As existing cotton fields exhausted and harvests diminished, planters cleared more land in an attempt to balance old ground with new, increasing the total area dedicated to cotton just to maintain the expected yearly product. The crux of the downward spiral, as expressed by a writer known as "Cotton," was the need "to increase yearly the quantity of land planted to produce an *average crop*—which crop gradually diminishes, as the land from continued cultivation becomes exhausted," until the planter clears still more land or "abandons his plantation in despair, and the result is—emigration."[118] Put slave labor into the logic and it becomes even more desperate. Fresh land produced more cotton per acre, so it allowed masters to employ slaves at a greater level of profit per slave; in other words, it increased labor productivity at no additional cost but the work of clearing. Plantations where cotton continually assaulted forested land could reach an extent, with slaves spread so thin, that there would be insufficient labor unless new slaves could be raised or purchased. Masters needed fresh soil not only to feed the exhaustion monster; they needed it to keep their slaves fully employed.

No one understood all of this better than planters themselves. As "Cotton" complained, "It is the system of planting largely to the hand which has so sadly impoverished the upper country of this State, is now wearing out the lands in the West, and will always produce the same results wherever practiced."[119] The same thing finally occurred to Law, his ambition tangled up in nut grass. He increased the number of his slaves into the 1850s as the number of his cotton acres also increased but with little to show for it. Decades later, in 1871, after war and emancipation, he spoke before the Hartsville Farmers' Club about the peculiar institution and the hard-run land it left behind: "In the past, while slavery existed amongst us, we were forced by the circumstances in which we were placed to enlarge our farms to afford employment and support to those entrusted by Providence to our care. . . . We all now have more cleared land than we can find reliable labor to cultivate."[120]

The violence of slavery begot the violent regard of slaveholders toward their land, but slavery may not have been entirely incompatible with improvement, and a very small number of planters went further than Law did to close the circle and eliminate clearing and long fallow. After all, explained still another writer in the *Southern Agriculturist*, an acre can be manured as easily as it can be cleared: "Everything depends upon the intelligence and industry of the individual who directs the labour; if he chooses to employ a part of [his slave labor force] in the making of manure, I should be glad to know what is to prevent him."[121] Stubborn, cussed unwillingness to change may explain as much as the inflexibility of slavery. Other things going on at the same time also account for the slope of South Carolina into stagnation by the end of the century. Rice from Bengal became part of a regular trade between Portuguese Malaysia and Europe beginning in 1843; rice from Louisiana, Arkansas, and Texas also cut severely into demand. With about 50 percent of the total wealth invested in slaves at the time of the Civil War, with a thinning population and low density in its cities, with little waterpower to compare with Massachusetts or the Delaware River valley, it is no wonder that South Carolina fell into decline.[122] Improvers may have desired better for their society, but they could do no better without bringing into question the planting system itself and all of its ferocious racial and social relationships. Conditions could be even worse when extensive cotton colonized environments that suffered under its rule. The erosion of the Piedmont suggests another connection between the agricultural ecology of South Carolina and the problem of permanence.

Cultivation amplifies the erodibility of soils, causing movement in a short time that would otherwise require centuries. Take off the scrub and grass, and the earth flows.

Thomas Law's journal is electric with the tension between attempts at good practice and an adverse environment for it. The dismay Law experienced almost every year during the 1840s came from rainwash. March 1–9, 1841: "Rained all day and also shelled corn and raked up manure. . . . Began to haul manure from lot, a few rain days until 9th, when it rained all night and incessantly until 12 at night [—] more water on the earth than I have ever seen, at one time in living on the place, the greater part of lot manure hauled in the field and exposed to all the rains."[123] Dung washed and weakened under rain and did little good thereafter. March 11–15: "Since the 9th 3 days and nights it has rained incessantly, never saw as much rain fall at one time, the whole country inundated with water." March 16 and 17: "Rain . . . and the ground already so wet that it can not be ploughed have opened all the ditches to let off the water draining almost nothing." A break in the weather allowed Law to plant the Gin House Field during the first week of April, every day directing his slaves to mix dung before plowing. Then it rained at a critical moment for the seedlings, on April 29: "Very hard and extremely windy all night and during the day of [the] 30th, this day moulding corn in mill field . . . cotton looks miserably what is up and corn very small." He lost an entire crop of cotton the same way in 1843, when it rained during the September harvest: "Half sick . . . and in a melancholy mood and not much prospect of a *change for the better*." In July 1847 he again recorded "the greatest fall of rain ever known in these parts . . . every sink filled with water, out to the hills more [than] 20 acres of cotton drowned and about as many of corn . . . —crop ruined beyond human remedy, my own loss can not be less than $500." Then two years later came "The Great rain of 10 July 1849." It went on.

Law hardly mentioned lost topsoil from erosion. Had he attempted cotton even on a shallow incline, he could have measured the loss in cubic tons. Instead, he lived in a place where deep sand accumulated near river bottoms to form rich places for crops.[124] The remarkable intensity of rain on exposed ground through spring and summer resulted in other kinds of loss. It leached nutrients to the subsoil and caused cracking, hardpan, and sheet erosion—forms of attrition that often took place below notice. Like most American farmers, Law folded all of these into run-of-the-mill "exhaustion," where it escaped accurate accounting. Just north of Darlington the terrain rises and changes as it becomes the Piedmont, a drastically different geology where the environment presented a fearsome obstacle to livelihood and settlement.

David Golightly Harris worked the Piedmont. A middling farmer and slaveholder of the Spartanburg District near the Blue Ridge and the son of an affluent planter, Harris aspired to his father's status but spent decades hacking out a turbulent agreement with a soaked, sloping red-clay soil. In 1850 just 31 percent of the South Carolina Piedmont could be called settled, and that figure included land in grazing and fallow. So small an extent just ten years before the Civil War, by which time it had increased to 33 percent, suggests that the Piedmont had been perpetually broken for cultivation and then abandoned as marginal land. Writing in 1928, Hugh Hammond Bennett, who conducted some of the first government-sponsored soil surveys, reported on ecological morbidity in the Piedmont: "A single county in the southern part of the Piedmont . . . has been permanently ruined by erosion. The whole area has been dissected by gullies, and bedrock is exposed in thousands of places. . . . The land has been so devastated that it can not be reclaimed to cultivation until centuries of rock decay have restored the soil."[125] If southern emigration had an epicenter, it was this delicate region, and David Golightly Harris's hardship and tenacity demonstrate the ecological limits of improvement.

Any Piedmont planter could have knelt down and milled the trouble between thumb and forefinger. Stretching from central Alabama through most of North and South Carolina to northern Virginia, the southern Piedmont never took well to American food production, and by the first decade of the nineteenth century portions of it had begun to give out. A long plateau of soft and weathered residual soils with frequent hills and sloping ground, the southern Piedmont rises from two hundred to fifteen hundred feet and includes examples of the Cecil, Helena, and Appling soil series—each highly erodible. Some of this land is so pliant it has come to be known as "sugar" soil, for its tendency to melt and gully under rainfall.[126] And it does rain. The Piedmont is a place of driving showers during the growing season—from break of spring to early fall. The water picks up silt from creeks and branches as it gushes down the incline through the Sand Hills to the plains and delivers this cargo to tidewater rice plantations in what amounts to a transfer of topsoil from the high places to the low. Lying between the Blue Ridge and the coastal plain, the region captures a large share of the moisture moving north from the Caribbean. No settled place on the continent lost its tillable surface as completely as the southern Piedmont. That part of it lying within South Carolina became one of the most drastically eroded regions in the United States, a reputation it had already earned by the 1840s. The evidence forms a time line spanning half a century.

European settlement in the Piedmont dates to between 1740 and 1760, when a population of mostly Scotch-Irish farmers from Pennsylvania and Maryland bypassed Virginia and North Carolina and entered an isolated region where they planted subsistence crops, hunted abundant game, and enjoyed a feral existence that brought them into conflict with the affluent planters of the coastal zone. By the 1780s they had begun to plant cotton, and once the gin became available, the green-seeded staple spread throughout the backcountry. Slaveholding increased by 83 percent between 1810 and 1820.[127] A state atlas commissioned in 1825 refers to "Poor Land" and "Extreme Poor and Level Land" in the Spartanburg District, located high to the west and close to the mountains. The chief surveyor, Robert Mills, wrote of the Piedmont lying within the state: "This region is hilly, and in many parts too rolling for cultivation, without washing."[128] Twenty-five years later the same upland region had experienced relentless destruction. "Water-worn, gullied old fields everywhere meet the eye," said one writer; said another, "The destroying angel has visited these once fair forests and limpid streams . . . everything everywhere betrays improvident and reckless management."[129] A correspondent from New York put the pieces together in what may serve as a summary of the South Carolina soils crisis: "Frequent use of the plow, and the unremitted culture of the soil in corn and cotton, have not only deteriorated the quality of the land, but exposed the surface to be washed away by the heavy rains of these latitudes, and the traveler . . . encounters little else than bare hills of red clay, washed into hideous gullies or barren fields, overgrown with broom grass and low pines." Farmers cleared new land until they moved out, "so that the effect of this wretched system was not only to destroy the fertility of the lands, but still further to impoverish the State by promoting emigration."[130]

The United States Soil Conservation Service provides more contemporary evidence that something appalling happened in the Piedmont during the nineteenth century. The Newberry District is a thoroughly dissected plain, gently sloping to moderately steep, located in the watershed formed by the Saluda and Broad Rivers—a geology fairly typical of the Piedmont. In the 1950s, 68 percent of the county was covered by second-growth forest, or, as the service explained, "many large areas were abandoned after they became eroded and have reverted to forest that consists chiefly of pine trees."[131] Said the service, soil in Newberry might supply all the vegetable and cash crops known to the country and give any settler a good living, but listen to this: at any greater slope than 5 percent it needs to be terraced and cropped in long contour strips "to prevent serious erosion." Portions of the district had lost *all* of their topsoil by the 1950s, exposing the red rock below. Codes written in combinations of numbers and letters on Soil Service maps translate local knowledge into legible description: CdB2 (Cecil sandy loam, eroded gently sloping phase), AcD2 (Appling sandy loam, eroded strongly sloping phase). Eroded soils carry a 2 or 3, the higher number indicating more severe damage.

In one aerial photograph selected at random, land bordering the Union District and the Enoree River is pocked with bare spots and littered with 2s and 3s from north to south. A rash of 2s and 3s plague Newberry and Oconee and Spartanburg and Fairfield. A traveler moving from the nearly level, sandy clay loams of the coastal plain to the gently sloping, well-drained silt of the uplands would see things worsen on the way up. In Bamberg, one county away from the Atlantic, there is still considerable erosion but not nearly as much as in the Piedmont. A study conducted by the Southern Regional Committee of the Social Science Research Council in 1936 found that 61 percent of the nation's 150 million acres of eroded land (or 91.5 million acres) could be found in the South and that "a single county in the South Carolina Piedmont had actually lost by erosion 277,000 acres of land from cultivation."[132] It is not clear how much of this damage happened after land had been abandoned, but there can be no question that agriculture provoked it. Farmers left, trees came back. Cotton and corn once grew on thousands of plantations across what is now the Sumter National Forest. Loblolly pine, a weedy species especially common as second growth on old agricultural land, became a symbol of abandonment.

Geography and geology have a third partner in climate, and when combined they account for the earth-moving potential of the Piedmont. The force of precipitation on soil is quantified by the rainfall and runoff factor—known as the *R* factor. It is an index computed by adding the total kinetic energy of annual rainfall in a given area to average precipitation during the thirty-minute period of greatest intensity.[133] The southern states have *R* factors that range from 350 in south-central Georgia, to 400 on the coast of South Carolina, to over 500 at the mouth of the Mississippi River. By comparison, New England ranges between 75 and 150, while Minnesota and Iowa never exceed 175. Annual rainfall, however, reveals little or nothing about these differences. Precipitation from Maine to Georgia stays within a fairly tight range of thirty-two to forty-eight inches. It is that other measure—the maximum intensity recorded during one half-hour period—that better explains sharp regional variation in the *R* factor and why cultivation in the South Carolina backcountry could be so dangerous. One and a half inches in one hour can pummel South Carolina in the early spring, with fields plowed up, planted, and vulnerable. Keep in mind that raindrops pack two hundred times the kinetic energy of surface flow and do much more damage. In the words of one investigator, the detaching of soil particles—as opposed to their transportation in runoff—"is by far the most important part of the process." Along these lines, frozen ground provides another reason for the dramatic variation between North and South. When spring rain falls on snow or ice, it dislodges little soil. Finally, if there is any doubt that the quality of rain is more important than its quantity when considering the erosive power of water, consider the case of the Pacific Northwest. It rains virtually every day there, but mostly in drizzles and fogs that do little harm to soils. The region's wettest hour in a year delivers just half an inch, which translates into an *R* factor of only 35 to 50 in the cultivated valleys of Washington and Oregon.[134]

A cropping system could make a great difference in the fortunes of a region, it could build organic matter against erosion, it could be highly adaptable to climate, but it could never alter the geological foundations of land. Even with the best management, it is not at all clear that agriculture in a region of such extreme disadvantage as the southern Piedmont would ever have been profitable. The administration of the Patent Office learned all about these disparities when it sent a circular to rural counties in 1848. One response came from Delaware County, a patch of 113,280 acres fourteen miles south of Philadelphia. The respondent claimed a total product from crops alone (not including livestock) of $1.7 million during a drought year, including 24,400 tons of "stray and corn fodder," 38,250 tons of hay, and 3,000 bushels of clover and timothy seed. He went on for eighteen pages. The one-page response from Dunlapville, South Carolina, located in the Piedmont, could not have been more contrasting. All injury and loss, the writer hoped for some future restitution: "I have no doubt that in a short time a more proper diversity of employment will take the place of the present too exclusive attention to cotton, and that our favored region will again be prosperous." Another respondent from the same state and vicinity did not bother with the survey at all but instead copied intimate evidence from his own plantation journal: "High, poor ridges, in amount many thousand acres, did not yield more than a peck of corn. . . . the whole vegetable world seems to wither under the curse of an angry Deity."[135]

David Harris knew the fickle ways of the same god. Most of all he dreaded the "hard, beating, washing rain" that immersed his farm throughout the late summer and fall of 1859. With ground saturated and exposed for plowing in February, he worried because "all night and all day it has been raining constantly. Every place and every thing is floating in water." On fair days his slaves picked cotton in the mud and took corn from the hillsides, where Harris often directed them to open clogged ditches.[136] The ditches salvaged sliding debris so it could be shoveled back into place and create level terraces for planting. Harris "gathered fodder" for his cattle by cutting feral grasses for hay and by raking stubble from his harvested corn. He also sent his cows to graze in the woods, though he complained often about poor pickings ("Cattle still requires [*sic*] to be fed and my cow-food is nearly out"). He spread manure, but only for wheat and the garden; and he frequently cleared land for crops.[137] Manure can counteract the effects of erosion by increasing the water retention of topsoil and by adding structure to loose, sandy loams, but the quantity necessary to achieve this effect would have been far beyond Harris's capacity to provide.[138] Like the Darlington planter, Harris made "newground" almost every year and fell behind the cycle, cutting the forest to make up for land he could not renew—but with fewer resources to buffer him.[139]

In spite of his attempts to stay the effects of hard rains on soft earth, Harris's fields showed signs of injury. He revealed their threatened condition in 1856, when he spent November "spading & turning up some of my bottom where it is poor on top but is very rich under side." Later he described digging to a depth of three feet and finding what he believed to be good soil after removing eighteen inches of sand. He hoped to improve his other fields by hauling some of this "rich dirt & put it on some poor spot & see if it will pay." It appears that he unearthed a layer of humus under recently transported sand. Whatever it was, Harris sensed the futility of the work: "I am making a good piece of land out of poor, but I am afraid that it will not pay." The same spring he complained of land "so crusty that I have again harrowed on my cotton that it can come up better."[140] The crust formed when loosened particles of soil floated to a new location and dried, making a surface impermeable to water and obstructing to seedlings. Moving good soil from one place to another sends up signal flares. Harris set up an ecological triage in which he prioritized places on his farm according to need. An observer of these conditions described a dismal improvement, what Harris must have understood as a desperate assault against a soaked and gravelly poverty: "When the soil has been washed away, and the underlying strata laid bare, the question of reclamation becomes one of formation." This was the burden of the Piedmont planter, and Harris seems to have carried it to his grave.

What the Agricultural Society of South Carolina said of the modest planter in bad times applied directly to David Harris: "Without a reformation, now that the virgin soil is nearly exhausted, the middle classes of planters must inevitably, in a few more years, become poor." Harris enjoyed no luxuries and seems to have owned nothing but his land, his A-frame house, and his few slaves. He must have felt beggared and eaten out when he couldn't scrape up enough bread to feed his children at school. Poor from poor land, Harris represented the typical candidate for emigration. What made him different, and made possible his continuance, were his modest attempts to conserve the basic resources of his land. Yet Harris is important not only because he stayed in Spartanburg through the Civil War but because he *believed* in staying. Most who espoused and practiced some variation of the restorative system had capital and standing, but not all. His aging parents and a number of local creditors kept Harris from leaving, but he also expressed a desire for continuity. Planting peach trees one winter day, he reflected on future times: "I hope that my children reap the reward of our labor . . . for it is the duty of one generation to work for them that is to follow and to whom we must give place."[141]

Hundreds of thousands of South Carolina planters did not give place to the next generation. The success of Law's swidden and Harris's triage belie a more general failure. Many did not have enough land for a long fallow, did not expend the labor to terrace their fields, did not strive for stability. When the first glow rubbed off and their countryside began to grow old, planters incapable of restoration could not resist the not-so-distant territory that, so they told themselves, would never need restoration.[142] Slaves transported to more fertile ground earned higher returns—a skimmer's efficiency. Alabama and Georgia had good access to the Atlantic and the Gulf of Mexico. Louisiana offered trade up the Mississippi River. The Old Southwest, recently "abandoned" by the Choctaw, Cherokee, and other tribes after Andrew Jackson removed them to Indian Territory, finally lay as empty of inhabitants and tribal rights as whites had always assumed it to be.

"Our Most Fatal Loss"

Politics, disease, curiosity—emigration cannot be reduced to a single cause. Yet planters consistently pointed to squandering soil practices leading to a decline in the productive power of land as the wheel of removal. The nervous sarcasm of one correspondent to the *Cultivator* pointed to the near impossibility of reform and the intractable cycle of land killing. "We Virginia farmers," he begins, "(I mean such as I am, who are at least four-fifths of the whole,) require to have some plan devised, by which, without *much labor* and with *no expense*, we may improve our lands, and that speedily, or we will remove to the western forests, and encounter all the labor and privations attending a new settlement. We have no notion of submitting to the tardy and laborious system of your *real farmer*. We go for a kind of *sleight of hand* or *no work* plan—or we are off." The course he followed was to plant corn one year and "rest" the land in wheat the next "and so on, until they are prepared for a good crop of *old field pines*." He concluded that if he could find a way to save his land for nothing, "it may enable *my class of farmers* to remain in the Ancient Dominion; otherwise (unless indeed you can reclaim us and our *lands too*) we must remove."[143] The choice could not have been stated more clearly. Edmund Ruffin acknowledged the satire when he published it in the *Farmers' Register* but commented stoically that "we know of no plan by which improvements may be made and profits gained, without labor and expense." Real farmers spent the money, expended the labor, and stayed; the other class could not be reclaimed. In South Carolina, oldfield pines always bent toward Alabama.

Thomas Law went to Alabama in 1834 not to stay but to look. It's difficult to imagine how he could have resisted, since so many of his neighbors and friends had gone. Law thought of doing the same, beginning with a visit to the abundance frontier, traveling with other "land hunters" to a territory he knew just two things about: "only been acquired from the Indians a few years but very fast improving." He visited towns he called rough and ugly, thrown up in a year or a day and already "in a state of decay." He heard a preacher from South Carolina get going a revival to rid Demopolis ("very sickly") of disease. He camped in the Cane Breaks after falling in with a Choctaw man. Together they navigated spiky reeds and swamp streams, then found a safe place to sleep in the still of the bog—soaked and mud-splattered, phosphorus marsh glow all around them.[144] Next morning Law went his own way to relatives and found the rumored land he came looking for—gray, very fine, sandy down eighteen inches. The product of it made him swoon with the blush of a new country: "The crops on these lands surpass anything I ever beheld." He returned home in early October with uncertain intentions. It was the first of two trips he would make, though he wrote little about them. Years later, at the age of sixty-one, Law talked about having never left Darlington. In an address before the Hartsville Farmers' Club he called the ceaseless clearing of the last remaining woodland in the region "suicidal" and concluded that no place on the cultivated earth is really better than any other as long as soil is accorded respect. He had learned by staying what many others never learned by emigrating: "Even the rich Cane Break Lands of Alabama are said to be failing, from being overtaxed by continual culture, and no return of vegetable matter to the soil."[145]

Another planter, Edward Means, made a series of trips to Alabama and Louisiana during 1846–1847 looking for resettlement. Writing excitedly in November from west of Montgomery, he told his wife, Claudia, "It is the most beautiful and charismatic county to live in I have ever seen[;] most of the residences are on high sandy hills breaking off into fine rich level lands for planting. I have been to see several plantations, which are for sale, 2 of which I would much like to own, about 15 miles from Demopolis and the same distance from Eutaw, one improved with a good house, the other which is directly opposite 1/2 mile with no house except one for the overseer." These plantations were only a few miles from relatives and friends who had made the same move earlier and who would have eased the transition. But

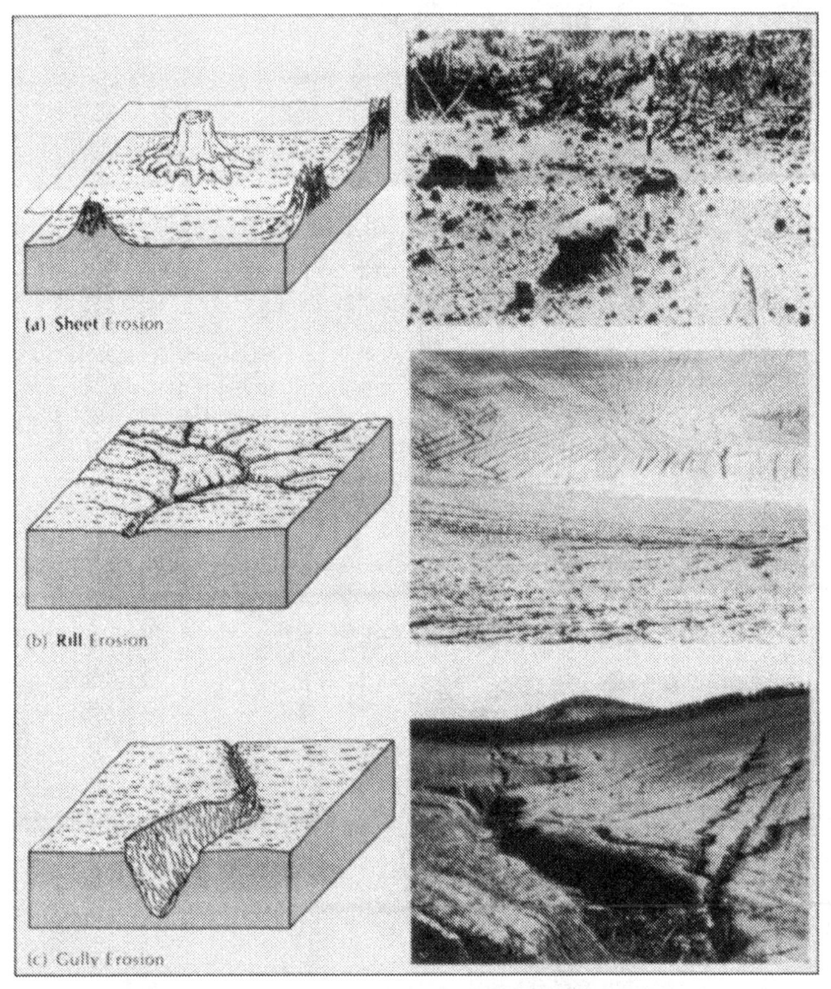

Erosion as exhaustion. American lands showed the signs of neglect in the form of wind and water erosion. Where soil fell away in sheets (evenly across a cultivated field), farmers made no distinction between attrition and infertility. Reprinted, by permission, from Nyle C. Brady, *The Nature and Properties of Soils*, 10th ed. (1990)

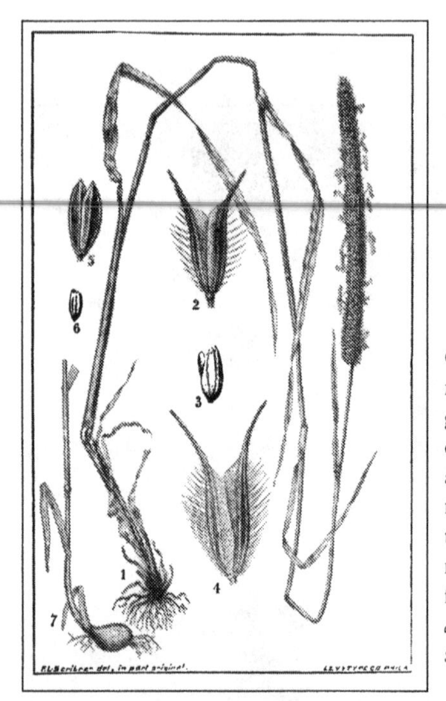

Grass and cattle husbandry. The combination of timothy and other nutritious grasses with well-bred dairy cattle not only created lucrative commodities but also created high-quality dung for the restoration of soils. This "perpetual system" required farmers to dedicate a significant portion of their land to animal feed. W. J. Beal, *Grasses of North America for Farmers and Students* (1887) and *Transactions of the New York State Agricultural Society* (1851)

Whether they learned about it from books, paintings, or their own travels, northern improvers fantasized about the English countryside. When English travelers found the American landscape wanting, they had places like Wollaton Hall and Park in mind. Jan Siberechts, *Wollaton Hall and Park, Nottinghamshire* (1697), Yale Center for British Art, Paul Mellon Collection

Apex of the northern farm. Edward Hicks's depiction of Bucks County, Pennsylvania, captures the richness, diversity, and labor intensity of the improvised farmstead. Edward Hicks, *The Cornell Farm* (1848), National Gallery of Art

Progress of a homestead (*above and opposite*). These panels tell the story of a farm at four stages: clearing, first summer, ten years later, and forty-five years later. They depict landscape evolution the way northern improvers imagined it, with successive owners investing increasingly in the same property until, as the author describes the final panel, "the scene has progressed to a consummation!" O. Turner, *Pioneer History of the Holland Purchase of Western New York* (1850)

Jesse Buel. No other improver attained his intellectual acuity or ethical vision. He applied Whig political principles of obligation, legacy, and progress to the condition of farmland and presaged conservationist thought. Yale University Library

Slash and burn. This form of clearing, also called swidden, quickly transferred nutrient elements from biological systems to soils and saved labor. Backwoods settlers and southern planters used it to take possession of new lands. Note the clearing in the distance. George Harvey, *Harvey's Scenes of the Primitive Forest of America* (London, 1841)

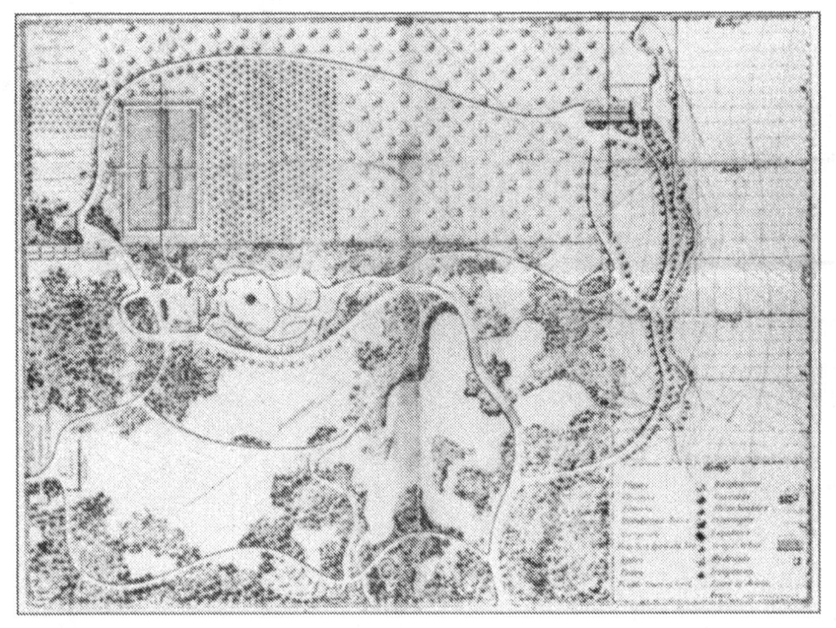

A country place of sixty acres. The aesthetic of neatness and order that characterized the improved farm of the 1820s had become a desire for beauty by the 1850s. This estate reserved twenty-eight acres for flowers, gardens, buildings, lawns, woods, ponds, and roads. Robert Morris Copeland, *Country Life: A Handbook of Agriculture, Horticulture, and Landscape Gardening* (1866)

George Perkins Marsh as a young man. Marsh is the consummate bridge figure between old-state improvement and conservation. Special Collections, University of Vermont Library

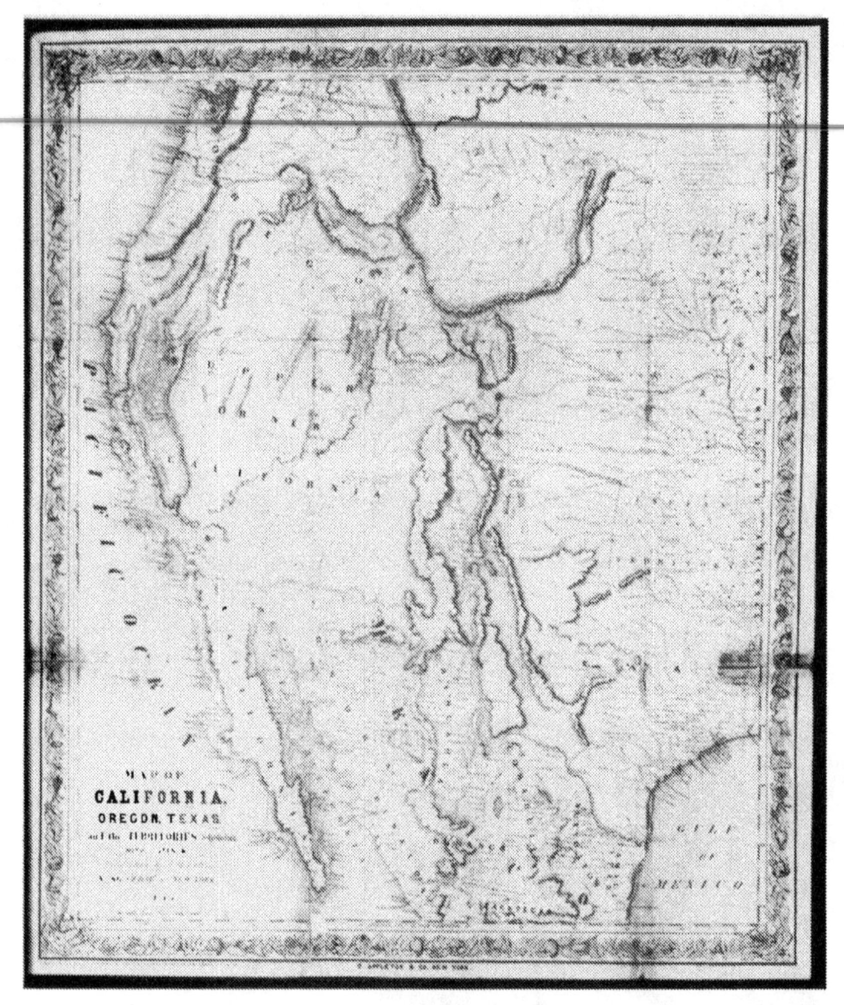

Continental America, 1849. A typical map celebrating the American conquest of North America and advertising known trails to the Pacific. It traces the route taken by General Stephen Watts Kearny along the Arkansas River to New Mexico during the Mexican War, as well as the Santa Fe Trail. The goldfields of the northern California Sierra—the only destination that mattered to overland emigrants—are exaggerated in size. Yale University Library the Means family never left home. Edward died of an unknown illness in New Orleans before his return trip home.[146]

The contagion of emigration reached high and low. James Henry Hammond—born in the Newberry District, a furious defender of southern nationalism and slavery, and governor of South Carolina from 1842 to 1844—felt the same temptation but held off long enough to attempt one last push on his worn land: "If I can get 800 or 1000 acres marled [with calcium carbonate to lower the acidity of the soil] . . . I shall be contented here. Two years will fully test this experiment, and if it fails I have no hope left but emigration." Hammond claimed that hardly a single one of the young men from his youth remained in the state. He purchased land in Texas but never moved to it, reluctant, according to his biographer, to leave "his political ventures and personal ties."[147] In 1844 he tried to redirect the efforts of planters, if not toward improvement then at least toward more profitable crops, as a way of keeping them from the territories. Hammond said more—that the era of cotton in the uplands of South Carolina was over and that planters would soon be compelled "to abandon its longer cultivation." Those who refused to listen would have no choice but to leave for where cotton still made a profit. He could only hope that their numbers remained small and expressed confidence that "there are some—a large majority I trust—who prefer to link their destiny with the land that gave them birth and struggle on at home through all the changes which time and fate may bring."[148] Hammond struggled on at home and ended up a bystander to the aftermath of emigration.

Many others left for good and never looked back. The collision between land and practice caused an out-migration that left the established planters of South Carolina trembling. *Niles' Register* reported on the erosion of society, describing it like an army falling back. Reported from Charleston, Virginia: "The tide of emigration through this place is rapid, and we believe, unprecedented. It is believed that not less than 8,000 individuals, since the 1st of September last, have passed on this route. They are principally from the lower part of this state and South Carolina, bound for Indiana, Illinois, and Michigan—They jog on, careless of the varying climate, and apparently without regret for the friends and the country they leave behind, seeking forests to fell, and a new country to settle." As the Nullification Crisis came to a head, it caused some white unionists to flee the state, or, according to a writer for the *Portland Evening Advertiser* in Montgomery, Alabama, "It would seem as if North and South Carolina were pouring forth their population in swarms. Perhaps I have gone by in the Creek nation over three thousand persons, all emigrating, including Negroes of course. The fires of their encampments made the woods blaze in all directions. . . . Politics in South Carolina have had much to do in accelerating this emigration." A correspondent to the *Farmers' Register* also blamed the cold war between Columbia and Washington for the mood of depression and the lack of exertion: "We prefer moving to the virgin soil of the west, to restoring by proper and judicious culture, the land given to us by our ancestors." Nullification was clearly an excuse or a final nudge for a larger motive, because even as late as 1854 the *New-York Tribune* reported, "No immigration whatever takes place to South Carolina, while emigration is so common, that even Colonel Calhoun, the son of the Statesman, left Carolina to go to Alabama to acquire wealth!"[149]

The number of people who moved was remarkable. A census of the white population taken in 1829 showed an increase of only 5.5 percent since 1820, representing a gain of 13,533 people. The state continued to grow between 1830 and 1840, if a rate of 0.47 percent can be called growth. Considering the range of statistical error, population may actually have declined.[150] The Patent Office estimated that between 1840 and 1848—the peak years of emigration—residents declined by more than 25,000, or by 3,200 people a year. The Census of 1850 tracked families by state of origin, and with that ledger it is possible to establish an estimate for the number of people who migrated from South Carolina to all the states of the Old Southwest, including Georgia, Alabama, Mississippi, and Louisiana. The census cannot tell the number who vacated and then died before 1850, no insignificant number considering that Alabama became a state in 1819. Nonetheless, by 1850 about 137,400 people had recently left South Carolina. The state's delegation to Congress shrank. Between 1830 and 1840 South Carolina sent nine elected officials to the

House of Representatives but only seven in 1850, then six in 1860, and fewer still after the Civil War. In 1870 only four representatives stood up for South Carolina, a command of only 1.6 percent of the House. As the number of western states increased, South Carolina held a smaller and smaller portion of the seats at the same time that its absolute numbers were falling.[151] It is never clear just what effect depopulation has on a region, since some people benefit by it. What is important is how the stayers responded, with writing and worry.

Here are voices from the great emigration, in which agriculture failed to safeguard political potency. In an essay on embankment and draining, an author lamented the borders like broken earthworks: "So much wealth; so many producers; so much strength—go from her borders never to return. Be this amount great or small, we have lost so much importance as a State; and to that extent the West is a gainer."[152] Another South Carolina planter dreaded the brain drain:

> Most of us have children, relatives and friends, who have left the state and gone westwardly, to seek for new lands. Many more, distinguished for talents and enterprise and public spirit, may be expected to follow, unless something can be done at home to afford them profitable occupation. The rank of South-Carolina among her sister states, will be diminished in proportion to her diminished population and productions.[153]

The 1840s brought the crisis to the boiling point, and many who had remained silent spit their frustration:

> What impoverishment, what ruin, what desolation has the spirit of emigration produced in South Carolina? . . . Look at the South-west and see there the outpouring of her citizens and her treasure! Look throughout the State and see their deserted fields and waste habitations! . . . No one considers himself permanently settled. No one expects his children to live where he does, to inhabit the house he does, or cultivate the soil which he is improving. . . . "If," said a once distinguished son of Carolina, "some desolating influence were to spread

over our country and sweep from it every human being, what vestige would there be a hundred years hence, of our civilization, or even of our existence as a people?" None at all.

The only lasting monuments to the previous occupancy would be "here and there a pile of rubbish."[154]

The loss became literature. Simple, affable Richard Hurdis breaks his mother's heart and flees engagement and smoldering sibling conflict to make his fortune in the newly wrested Indian lands: "The Choctaw territory was reported to be rich as cream; and I meditated to find out the best spots . . . as soon as the government could effect the treaty which should throw them into the market." As William Gilmore Simms moralizes in *A Tale of Alabama*, Richard deserts the lovely Mary Easterby for a romp in the West that convinces him of nothing else than that there is nothing to be gained by going there. Out from home just a few days Richard overtakes a band of emigrating families "from one of the poorest parts of North Carolina, bent to better their condition in the western valleys." They try to govern themselves exactly as before, to re-create the same society they had known, but Richard sees that "even this arrangement does not supply their loss, and the social moralist may well apprehend the deterioration of the graces of society in every desertion by a people of their ancient homes." A policy brief follows in which Richard decries the loss to civilization caused by too much land too easy to possess: "The wandering habits of our people are the great obstacles to their perfect civilization. These habits are encouraged by the cheapness of our public lands, and their constant exposure for sale." But while he teased their fascination with the backwoods, Simms never wanted his readers to forget that the places emigrants left derelict were the common places of social life: the deserted cottage covered with weeds, the Sunday church bell and schoolmaster now desolate. A people with no ethic of consumption will founder in good fortune, so "wealth, under such circumstances, becomes a curse."[155]

Simms fought the times. He railed against the desire for gain he saw pervading every aspect of American life, leading to moral and national destruction. Not really an improver but a "social moralist," Simms took the degradation of agriculture as the telltale sign of decline. Nothing exemplified greedy madness like staple-crop, market-directed agriculture: "The whole labor of the Planter was expended,—not in the cultivation of the soil . . . but in extorting by violence from its bosom, seed and stalk, alike, of the wealth which it contained. . . . A cultivation like this, by exhausting his land, left it valueless, and led to its abandonment."[156] Simms gathered all the data he needed for his novel by traveling in his own state—homes left to the wind and rain, evidence of a halfhearted nativity. The skimming planter laid out no gardens and planted no trees, "whose mellowing shade, covering the graves of father, mother or favorite child—would have seemed too sacred for desertion." "These, are the substantial marks of civilization, by which we distinguish an improving people."[157] He took his lament to its source and delivered these words at the University of Alabama in 1842. Yet something was missing from the moral. Simms never questioned slavery or the planting system that made southern families contemplate emigration. He asked for nothing more than a new habit of the heart with no attention to the underlying material problem; so, like a dead-leaf diet, his words carried little power to regenerate and left the prevailing assumptions and practices intact.

Society in South Carolina fell into decline after 1819 and suffered the effects of a mass exodus. Those who stayed behind carried on, and the state did not empty out, but South Carolina went into the Civil War with its cities stunted by an agro-economy more characteristic of a staple-oriented colonial society than of an independent nation. It utterly lacked the wherewithal to stimulate vigorous trade or manufacturing. Henry C. Carey, a Whig political economist and inveterate critic of the South, placed the dense and thriving countries of the world, including "Belgium, and England, and New England," in opposition to the poor and starving ones, including "Ireland, India, South Carolina, and Virginia." He then drew a chilling analogy. Disease-ridden Ireland, where a despairing population produced almost nothing, provided the chief European market for American surplus food: South Carolina did the same for Ohio. The reason had to do not with tariffs or abolitionists but with a dismal cultivation: "Nature has done for that State every thing that could be done; but man has, as yet, done nothing but exhaust the poor soils."[158] One Virginia planter staked his life's work on escaping this conclusion.

A Mouth Full of Ashes

Flecks of white, the bits and shards of seashells washed up on antediluvian beaches, now layered and crumbling near the surface of tillable land—in this stuff planters believed they had found a more perfect improvement. They knew it as marl, fossilized calcium carbonate that, like limestone, lowered the acidity of soils and enhanced the growth of plants. Marl tells another story of South Carolina, one impossible to disentangle from the complexities of convertible husbandry and the intransigence of plantation slavery, one impossible to disentangle from the life of Edmund Ruffin. Southern patriot, heir to John Taylor's agrarian severity, Ruffin left his Virginia plantation in 1843 and traveled to South Carolina, where he conducted a survey of lands over-tilled and drained away.[159] Marl became a tool in the service of Ruffin's grim and calcified quest for southern independence, in which the middle ground had disappeared and the world became reduced to stark choices: restoration or emigration, free or slave, union or secession. But not only did marl and the "Ruffin reforms" fail to stop the emigration and restore the tilth of cotton lands; their failure brought leaders to the conclusion that agriculture itself had failed in South Carolina.

This is getting ahead of the story, for the science and practice behind marl could not have been more sound. In fact, the discovery that soil pH can be regulated by the addition of calcium carbonate must be counted as one of the most important discoveries in the history of agriculture. Consider the problem. The great majority of southern soils are acidic. Heavy rainfall and the constant percolation of water remove base-forming cations, lowering pH to a point harmful to the typical variety of cultivated plants. Alfalfa, sugar beets, red clover, peas, cotton, timothy, barley, and wheat are known to be sensitive (in varying degrees) to low pH. Most plants reach their potential at a pH of between 6.8 and 7.5, or from nearly neutral to slightly alkaline.[160] The injuries that acidic soils inflict on crops are legion. They burn seeds and roots; they free heavy metals (like copper), which are poisonous at high levels; and by discouraging domesticated plants, they let the sun shine on native weeds and grasses, like the scourge that attacked Thomas Law's cotton. Not only does lime increase the absorption of soil nutrients; it adds one, calcium, which causes larger and denser crops—all the stronger when confronted by weeds and insects. Liming made some soils tillable for the first time, a boon to the South.

There was genius in it, but also error. One point stands above all others and provides the key to understanding what Ruffin did and did not do for South Carolina planters. Aside from supplying calcium, marl does not contribute critical nutrients. It does not build organic matter or counteract erosion. In soils with a neutral pH it does nothing at all. It is best considered an *amendment* and not a fertilizer, since it enables but does not create fertility. None of this qualified as mystical knowledge to improving farmers of the North, one of whom warned readers of the *Cultivator* that lime "does not possess any fertilizing principle in its own composition." Dress your fields once with it—dress them twice—but unless "every succeeding repetition be accompanied with ample additions of farm-yard manure . . . to supply the loss thus occasioned by the exhaustion of the vegetative power, every future crop will be diminished." The magic of marl was really a sleight of hand: by unlocking a store of nutrients in so-called sterile lands, it could be mistaken for fertility itself. Ruffin knew better than that; he knew the work of Humphry Davy (British) and Justus von Liebig (German). Davy asserted the "humus theory," essentially the scientific codification of organic or animal manures. Ruffin read Davy's *Elements of Agricultural Chemistry* (1813) when he was nineteen or twenty years old, and it led him to dig out a marl pit on his first plantation at Coggin's Point.[161]

Ruffin's reading of Liebig's *Chemistry in Its Application to Agriculture and Physiology* (first English translation, 1842) came later in life and confirmed his commitment to marl. On the surface of it, Liebig's book may appear conventional. He recommended building humus through animal manures and supplementing with legumes and grasses. Liebig, however, changed the terms of the discussion with the breathless argument that fertility could be broken down into a set of fundamental elements and that soil needed only chemical adjustment to remain highly productive. There could be increasing fertility only through additives to soil over and above what a farmer might apply from within his own resources. Or, as Liebig put it, "we could keep our fields in a constant state of fertility by replacing every year as much as we remove from them in the form of produce; but an increase of fertility, and consequent increase of crop, can only be obtained when we add more to them than we take away." Animal excrements could be replaced with "other substances containing their essential constituents." No one had ever said it that way before; although farmers had added all sorts of things to fields—seaweed, bird droppings, plaster of Paris, ashes—Liebig gazed across vaster waters than most experimenters to a time when the essential nutrient elements would be distilled into "solutions" or fertilizers. "A time will come when fields will be manured with a solution of glass . . . with the ashes of burnt straw, and with salts of phosphoric acid, prepared in chemical manufactories, exactly as at present medicines are given for fever and goitre."[162] If crops depleted inorganic nutrients as they grew, why bother with wet, smelly feces at all? Why not infuse the soil with nitrogen, phosphorus, calcium, and ammonia salts directly? In a series of experiments translated in 1855, Liebig tried to kill off the whole idea that plants needed decaying organic matter in order to grow: "The entire action and benefit of stable manures may be produced by use of mineral manures. . . . Organic matters, however useful they are in manures, may be dispensed with; art is capable of providing a substitute for them."[163] It is only a sliver of the professor's work, but it tells all. Sick soils needed medication administered in dosages and derived in laboratories. The cure could be abstracted from the messy agronomic process that created animal manure; it could be abstracted from the farm. In a modernist reduction with stunning implications for the future of rural environments all over the world, Liebig determined that the purpose of fertility was to feed plants, not maintain the flora and fauna of soils. He asserted the inorganic theory like a doctor prescribing vitamins for nutrition instead of food.

It may be possible to see Liebig figuring in Ruffin's thinking, but Ruffin came out with many of his ideas long before reading Liebig. More to the point, he never argued against humus or organic manure. What complicates Ruffin's career and marks his thought as a crucial southern synthesis, however, is that he did not go around exhorting planters to a grass-and-fodder husbandry. Without ever denying the efficacy of dung, he clearly demoted its importance in the minds of his readers. What exactly is he alleging in *An Essay on Calcareous Manures*? "There is one ingredient of which not the smallest proportion can be found in any of our poor soils, and which, wherever found, indicates a soil remarkable for natural and durable fertility. This is *calcareous earth*. These facts alone, if sustained, will go far to prove that *this earth is the cause of fertility, and the cure for barrenness*." On the one hand, the statement only reiterates the limiting tendency of highly acidic soils and the futility of manure wherever nutrients remained in chemical lockup; on the other hand, it also says that marl "is the cause of fertility."[164] To anyone unfamiliar with the man and his mission, the report Ruffin delivered to the state of South Carolina after the completion of his tour in 1843 looks bizarre. Just 13 pages out of 120 in this "agricultural survey" are dedicated to customs and errors in cultivation. The rest reads like a jumble of a dissertation on marl formations, the location of marls by watershed, Miocene marl, and post-Pliocene marl and includes all sorts of arcane information about shells, limestone beds, and burning lime. Ruffin preached and analyzed the practice and geography of marl in hundreds of writings through nearly forty-five years of his adulthood.[165]

Yet marl tells only part of Ruffin's story. He clearly understood that marl mostly prepared the ground for organic manures, and said so on a number of occasions. He complained of too much cotton and rice and not enough corn, grass, and provision crops. The lucrative staples might bring "great luxury and expenditure," but they would never guarantee the "plain and solid comforts" of a more diversified practice. He wrote that a system uniting livestock and hay "is essential to the permanent, solid, and agricultural prosperity of the State."[166]

In one of his last addresses on the subject he finally made himself clear: "When urging the use of lime, I have never omitted to state that it gave no fertility of itself, or by direct action; and that vegetable matter in sufficient quantity, and in conjunction, was essential to the beneficial operation of calcareous manures. The required organic matter may be supplied mainly in the growth of the land to be improved. But it *must* be supplied in some form, and in sufficient quantity."[167] Yet in the many volumes of his published work Ruffin rarely urged the complete system on planters. Without really spelling it out, he trusted planters to continue as they always had. He upheld the regional assumption that planting and farming belonged to different planets.

Ruffin's restoration—exemplified by marl itself—endorsed an open system, an extensive land-use regime. He said one thing over and over again with every particle in his body: the lands of the South contained all the elements of their restoration. On a tour of lower North Carolina just before the Civil War he paid close attention to the "composed manure" he saw in Edgecombe County. Every year at about the same time planters made a mash by selecting from a menu of ingredients: earth, marl, ashes, barnyard dung, cottonseed, salt, and guano (imported bird droppings). Ruffin saw one hundred mule carts of earth brought to every acre on one plantation, followed by thirty bushels of cottonseed. Slaves gathered dead trees and drift logs and burned them slowly to keep the ashes from flying. Of the 600 cleared acres on one plantation, 350 were covered with this compost, 50 more with bird guano. Ruffin declared that the improving planters of Edgecombe had invented a perpetual system: "Increased products and profits have been made on lands cropped almost every year . . . and without any thing like a rotation of crops. Cotton occupies the same ground almost continually, and always for at least four or five years in close succession." Planters could have it their way—no rotation and constant tillage without the cost and bother of grass or cattle. They could develop their own version of restoration to rival northern farmers. In fact, the Edgecombe compost mostly came from sources external to the plantation. It pinned fertility to exploitable land and so only replicated the problem it attempted to solve.[168]

This is the key to Ruffin's understanding of soil and the South. He faced a population stubbornly and inexplicably unwilling to change, a population that Hammond called guilty of "petty interests, false reasoning, unsound calculation, and perhaps above all, certain traditional habits of thought and action."[169] Planters showed no willingness to attempt the good system, and they reacted almost the same way to marl. Ruffin's mission as a reformer was to find a source of renewed fertility that placed only the most modest demands on planters and endorsed plantation methods as good and sound in the first place.

Ruffin's high profile as editor of the *Farmers' Register* made him the obvious choice to conduct a geological survey, and Governor James Henry Hammond invited him to South Carolina in 1842. Still, a number of vocal planters resisted the project as a waste of public funds. Yet after these planters lost or consented, expectations soared that Ruffin's conclusions would address everything from honeybees to poultry and stop emigration.[170] A strange polarity developed between Ruffin's public reception and his private thoughts. What the surveyor saw as he stepped out along the road left him heavy, pensive: "For much the greater part of the journey, the country appears like the former residence of a people who have all gone away." In every corner of the state he insisted that marl brought land back to life, though he wrote to himself that "not an acre has been tried, & an utter disregard or ignorance of the value of the manure is universal."[171] Nonetheless, a sense of excitement seemed to build as the survey progressed. He talked to the Agricultural Society of St. John's, Colleton, on Edisto Island on March 1, addressing a large audience curious to know all about marls, "how they operated—what lands they were adapted to—how they should be applied." Ruffin spoke from his seat rather than give a formal statement, and when he took out samples that he had picked up along the way, the display sent a charge of recognition through the room. "Several gentlemen, unaware of its great value at once declared that if that was an improver of soils, they could supply the island."[172] The questioning went on for two hours. Afterward, Whitemarsh B. Seabrook, president of the state agricultural society, read a letter previously published by Ruffin that rang the bell even louder: "Millions of dollars in value of newly created agricultural wealth will accrue to the state within a few years after the commencement of the general use of calcareous manures."[173] A spirit of industry, a reversal of decay, a rise in the enrollment at the College of William and Mary—all would be owed to marl. Said one planter from North Carolina, marling "had the effect of stopping emigration." Never had such a magnificent solution come from so common a source.

The discovery that a substance literally underfoot could turn sour land into sweet should have caused commotion. Marl offered planters all sorts of benefits at little or no expense but labor and transport. It could be applied for adequate results just once a year at a minimum quantity of two hundred bushels per acre, according to Ruffin. All the work to enrich sixty acres for one year could be accomplished by one man and a horse, and not every acre needed to be marled every year.[174] Compared with convertible husbandry, marl demanded less time and labor. To some it assured a new rural landscape by offering a stay of execution for the declining forest, in the way that all improvement translated into landscape conservation. One planter writing in 1841 announced a new ethic of consumption: "I am no enemy either in theory or in practice to the clearing of our forests—but I think the time has come when it should be done with great caution. Our upland forests of prime quality are nearly exhausted." Liming would propel staple-crop agronomy some distance at least along the gradient running from planting to farming by commencing "a system that promised really to give to the planter the character of a cultivator of the soil." Another believer estimated that marl would cause a threefold increase in corn per acre and a sixfold increase in profits, all on an investment of about $5 per acre.[175]

Then nothing happened. The underwhelming response is difficult to understand. The same writer who worried about the extent of the upland forests expressed amazement that planters sitting on top of the stuff worked in the same old ruts, even after the publication of Ruffin's *Essay on Calcareous Manures*, at a time when "South-Carolina was suffering severely from the emigration of her citizens." "Exhausted fields, lying over large beds of lime were deserted, in some instances they were literally abandoned, and at best, they were sold as lands hopelessly worn out, and their proprietors went to the great Western Valley to seek richer land."[176] James Henry Hammond's plantation was the exception that proved the rule. Ruffin's most dedicated supporter and diligent student, Hammond secured the rights to take marl from a place called Shell Bluff in November 1841. At great expense, he hauled it by boat to his plantation—eight hundred bushels per load, requiring eighteen slaves in six carts, in order to mix two hundred bushels per acre. "My expectation is to make all my marled land bring $225 to [$]250 average *permanently*."[177] The following year an article appeared in the *Farmers' Register* that was almost certainly written by Hammond. The author had the capacity to haul twenty-five thousand bushels of marl from a nearby bluff at a rate of twenty-four hundred bushels per week and intended to spread it on six to seven hundred acres for the next cotton crop.

Yet Hammond constantly revealed the conflicts and pressures of one trying to make old land pay like new without restorative husbandry. He made a telling comparison: "I think, as thinks every one, that the effect *this year* is fully equal to a fair coat of our stable manure." With the emigration of others and the possibility of his own always in mind, Hammond needed to prove that land well marled stood up to anything in Alabama: "I feel a strong desire to beat the Western planters on my *whole crop*, per *acre & per hand* and should think it done at a cheap rate if at no more cost than I should incur in moving, buying, settling & opening a place there."[178] Hammond's operation was exceptional in its scale and organization—far beyond what average planters attempted. Failing to beat the rich bottoms of the Mississippi valley, average planters were much more likely to remove.

Why did the crusade fizzle? Why did the solution to such a prevailing need receive so little response? Why did only a handful of planters, part of that minuscule number of ardent improvers, actually dig out marl and mix it in? For one thing, although soil pH reacts to the presence of marl almost immediately, it takes about a year before pH reaches neutral.[179] Planters probably gave up quickly. Said a northern improver about the want of immediate results, "This is the rock on which thousands fail. Lacking persevering patience, they become lukewarm in an enterprise—lose their interest, after which, total neglect, if not absolute disgust, ensues."[180] And though marl required no dedication of land, it did require labor. That layer of soil and sediment covering shell beds just below the surface could be five or ten feet deep, saturated with water, and covered with roots and trees. Once excavated, the marl might be thick as clay and very heavy to shovel and load. The peculiar ways that planters evaluated the difficulty of a task also had to do with what was familiar to them. "Many a planter who would shrink from the appalling labor of liming his land, will not hesitate, if it is convenient, to clear land"—maybe because they had been doing it for years. The soundest reason may have had to do with the *mentalité* of the masters, many of whom refused even to superintend their fields, leaving all decisions to overseers. The young men coming up after the generation of first settlers had a reputation for arrogance and dissipation: they would not scrimp and figure like Yankees, no matter the outcome. An offhanded cultivation was the only one they would suffer, because an aristocrat does not bend to pick up a penny even if it means not eating for a day.[181]

Yet nothing could have been as insidious to reform as the influence of slavery. As the historian William Mathew concludes, slavery stimulated reform as a source of protection around the institution itself, but it stood in the way of true reform by diverting the attention of planters toward the political conflict. Slavery "not only hindered much reform but was itself intrinsically unreformable" by the 1840s. Ruffin's marl program aimed at making slavery more productive, more profitable, more of what slave owners hoped it could be. Few if any tried to change slavery into something else. "As long as the dominant institution remained inviolable, generalized appeals for economic and sectional recovery were, at best, windy and ill thought-out and, at worst, specious and self-serving."[182] Lacking the will, as Ruffin himself accused them, planters continued as they always had as alternatives fell away. Slavery may not have been utterly incompatible with improvement, but it made any reform requiring the dedication of space and labor very unlikely.

By the time the editors of the *Southern Quarterly Review* returned to the subject of marl in 1849, they could only fulminate at a missed opportunity. The occasion for an article was the publication by the state geologist, M. Tuomey, of an illustrated *Report on the Geology of South Carolina* in 1848. Tuomey had published a report in 1844 as a coda to Ruffin's survey, but the editors responded to this one not so much for what it included as for what it said so loudly by omission. The editors wrote, "South Carolina, almost wholly an agricultural State, has not shown itself particularly anxious for the promotion of the peculiar interests of agriculture." Looking back on those days of promise when Ruffin walked the countryside, they lingered on what could have been: "We greatly regret that he failed to persuade our planters to the adoption of his Virginia practice. . . . They were impatient of the slow results." Planters used marl without experiment and ended up abandoning the whole project for lack of "reasonable time . . . and patient observation." Ruffin himself said the same thing: "I fear that nothing can induce them to marl *in proper manner*. . . . I have been grieved to hear that all my preaching on this subject has served to do but little good, even to my few converts." But he could not name the deeper cause of failure.[183]

Like long fallow, marl could be used as a way to renovate soil and save labor. That it did nothing to build organic matter, prevent erosion, or add nutrients would have no significance were it not for the fact that Ruffin almost never advocated a system that did these things in addition to lowering acidity. Since most "exhaustion" resulted from lost topsoil, so easy to let slip from land opened for cotton and washed over by rain, marl patched one wound while another one bled. Planters who excavated the stuff at the bottoms of gullies participated in a strange geological irony: restoration discovered in the very evidence of attrition.[184] By the end of the decade, at least a few prominent planters had begun to consider alternatives to agriculture itself as the basis of South Carolina's economy.

> At no period of our history, from the year 1781, has a greater gloom been cast over the agricultural prospects of South-Carolina, than at the present time.
>
> —*Southern Quarterly Review* (July 1845)

In November 1849 Hammond addressed the South Carolina Institute at its first annual fair.[185] The mood may have been festive, but he was not, and he used the occasion to enunciate fears and dead certainties in a long speech. For the majority of planters, cotton was no longer profitable, earning a bare 2 percent at one thousand pounds to the hand. "At such rates of income our state must soon become utterly impoverished, and of consequence wholly degraded." Capital had fled. Hammond estimated that $500,000 had passed with fleeing planters into other states. Slaves vanished, many sold west to great advantage, "and that institution [slavery] from which we have heretofore reaped the greatest benefits will be swept away." Improvement had failed. Neither the agricultural societies nor any of the available publications had "effected anything worth speaking of." Even the conditions that might have made improvement worthwhile seemed to have faded over the ridge, for it was a bad irony that it took money to make money: "Such improvements are never made but by a prosperous people, full of enterprise, and abounding in capital." Rich countries never depend on a single product like cotton, said Hammond. Since the short staple could be grown all over the South, his fellow countrymen spread all over the South to grow it. So South Carolina bled its white and black population: "Our most fatal loss, which exemplifies the decline of our agriculture, and the decay of our slave system, has been owing to emigration."

Then the speaker moved into new territory for an old republican and redefined the political economy of South Carolina with breathtaking abandon. Was not agriculture the sole occupation of a virtuous republic? Was it not a self-evident truth that the hard value in any nation derived from the productions of its soil? Hammond chucked the whole thing like an anvil around his neck: "These distinctions are only verbal—mere words without any philosophical or rational meaning." Wait a minute, Hammond! Manufacturing corrupted a nation by tying a free people to wages, the equal of political and economic dependence on merchant capital . . . right? "It is not true that these pursuits are hostile to political freedom. The truth is the reverse." Far from being the enemies of liberty, said Hammond, merchants, manufacturers, and lawyers "have always been the first 'to snuff tyranny in the tainted breeze,' and foremost in resisting it." As for the Tariff of Abominations and northern aggression to protect the products of its factories contrary to southern interests, "if the South manufactures for itself, the game is completely blocked." Manufacturing could launch a new southern ascendancy, and Hammond espoused factories as part of a theory of economic development. England became rich and great by its mill-driven industry and so would the South, he declared, as soon as its elites detached their slaves from agriculture, making them available for other kinds of production. Shackled to waterpower and the speeding leather belts that drove the new power looms, slave labor would stimulate a regional revival—people would stay and others would come. Of manufacturing he said, "The immense benefits to the South . . . would be so vast as to defy all previous calculation."[186]

Hammond was not alone in these views. Edmund Rhett, a resident of Beaufort, South Carolina, asked the members of his local agricultural society, "Who is the producer?" The great error of political economy in the United States, he asserted, stemmed from the French economists of the previous century "who maintained that Agriculture was the only source of all national wealth and revenue, and that three sets or classes of persons co-operated in the production of these—the proprietor of the soil, the laborer, and the merchant and artificer." Rhett and Hammond both took aim at the French Physiocrats and their doctrine—adopted by Jefferson and folded into the radical republicanism of the previous generation—that other classes only add to the value created by farmers. In fact, continued Rhett, "there is little praise in refuting an argument, which is no longer supported by any body. . . . Although all reject it as having no foundation in fact, it is common enough to see its hoary fallacies paraded before us in all the tidiness and pomp of the newest and most momentous truths." Rhett's own economy bristled with modernity and the detachment of wealth from any form of production. Wealth derived from a comparison between commodities in light of their relative demand; it came from the accumulation of exchangeable value. "It is the aptness of things to the wants and comforts of men, that gives one or another the superior value. Every result of human labor which accomplishes this, whether tangible or intangible, material or immaterial, no matter what, has of necessity some exchangeable value, and is so far an element of wealth."[187] By killing off the idea that all wealth derived from agriculture, along with the once-sacred political doctrine rooted in that premise, Hammond and Rhett found a way to escape the unattainable standard of convertible husbandry. Perhaps they had noticed that successful factory production did not require a system of nutrient recycling.

Hammond could not seriously have meant that agriculture would no longer be the mainstay of the economy, just as he gave too much away when he asserted that cotton would move out of South Carolina for good. He had a knack for dismal talk. Even in overstatement, though, he said something important. Agriculture would no longer, must no longer, be South Carolina's only livelihood. The same 1849 address contains another announcement, a kind of invitation. South Carolina had natural resources other than soil still to be found and taken: "We have coal and iron. We have . . . immense forests and noble streams without number. We have capital and labor, and the raw material is peculiarly ours. It only remains for us to prove to the world, that we have the courage to claim our own." Planters who abandoned their oldfield never realized the fine prospects they owned but could not see, and in 1849 everyone knew how unseen prospects could change the face of a country. Hammond knew about recent discoveries in the Far West and in a moment of political acrimony wrote privately that "sometimes I entertain serious ideas of removing to California."[188] So he must have considered that if found metal could make a once-desolate territory populous, then maybe it could save a once-populated state from desolation. It was a remarkable capitulation. Hammond traded away the idea that a countryside ought to be a place of sustained wealth from nature. By endorsing iron, coal, and trees as exploitable resources, as substitutes for the soil he once tried to conserve, Hammond called after all those young men cutting loose for overland trails, telling them, in effect, that the frontier is here! Come and take it if you have the courage to claim what is yours! This from one of the South's only true improvers, Edmund Ruffin's colleague, John Taylor's legacy.

The 1840s marked the end of an era. Territorial expansion, that national ship of progress, reached its perfection in James K. Polk's dirty little war with Mexico before running aground on sectionalism. In 1846, soon after the outbreak of the war, a Democratic representative from Pennsylvania named David Wilmot introduced an amendment to a military appropriations bill that became known as the Wilmot Proviso. Repeatedly passed by the House but never by the Senate, the proviso would have prohibited slavery in any territory acquired from Mexico. Nothing so deadly to the party system had ever come before Congress. It threw the entire weight of slavery politics into the question of territorial expansion and caused lawmakers to fall into ranks by section. If antebellum politics seems like a nineteenth-century version of Cold War domino theory before the Wilmot Proviso, it seems like the Cuban missile crisis afterward. Rather than calculate losses due to emigration, as they had done in the past, southern politicians came to regard mass removal as the sowing of a different kind of seed—a Greater South. The proviso fettered southern territorial ambition; insulted southern honor; and imposed an illegal law by disregarding the Constitution's own allowance of slavery. The possibility of no one knew how many western states carved out of the lands taken from Mexico made the geographical expansion of slavery a political necessity. Although that promotion may never have been the official policy of southern politicians, they began to assert it as a right like never before, and so it became part of the political combat leading to the Civil War.

Both sides stood ready to play the "Texas game" in the Far West, and everyone knew its rules: those who infiltrated a territory in such numbers that they composed the majority of its population would devise its state constitution. One writer in the *Southern Quarterly Review* mourned the loss of California, the slave state that never was: "We are shut out, forever, from access to the Pacific shore. . . . Can any one doubt the amount of emigration which would have wended its way from the South? By this time, thousands of young, intelligent, active men . . . would have been in that region, having each carried with them from one to five or ten slaves." In this context, old-state improvement had no meaning and no suasion. Every opportunity to filibuster became a crucial opportunity to extend a rigid conception of republican society, finally just a vacant mask for slavocracy.[189]

One filibuster captivated the popular mind and weakened every argument for improvement, a filibuster like no other. When, in December 1848, eastern newspapers printed Polk's confirmation that gold had been found in a California river, men began to leave towns and farms. The Gold Rush reached across class lines and appealed to respectable middle-class merchants and rising farmers just as it appealed to the indigent laborers who could not readily afford the price of passage. The very idea of self-improvement—once firmly tied to family, community, and the solemn obligations of property—went through a stunning reinterpretation when it became a justification for get-rich adventuring.[190] The idea that soil could be made to produce in the same place without end did not vanish from the rural press, and societies continued to form and fade, leaving their proceedings and memoirs like dinosaur footprints in silt, but improvement had none of its old urgency or potency in the old states after 1849. The Gold Rush was the ultimate removal.

The frontier never really lay over the ridge or across the prairie. As a borderland of ethnic conflict and insecure authority, it occupied a definable space and could be located on any map of the 1840s. As a place where people extracted all sorts of material from land at a profit and discarded the leftover, it could be anywhere at all. Though the state geologist Tuomey complained of the difficulty of separating "an Agricultural from a Geological Survey," he managed to do just that.

His report of 1848 consists of academic investigations—on stratification, paleontology, and mineral content. A chapter titled "Economical Geology" discusses manures and crop rotation, yet two other chapters stand out. First, Tuomey gave special consideration to each of the Piedmont districts. Manganese in Edgefield looked promising, so did mica slate in Abbeville, so did iron in Spartanburg. Expressing no interest even in soil composition, Tuomey seemed to be searching for some alternative extractive economy to replace food and fiber production in a region where they were no longer profitable. Tuomey also describes every known deposit of gold. The geologist found two hundred men working a mine where the Census of 1840 found just fifty-one eight years earlier. Another site had been overrun by miners, who apparently paid for the right to work a section twelve feet square in companies of from three to six men, "who work it as they think proper, and abandon it when they please." Tuomey went so far as to sketch, explain, and recommend various machines for extracting gold from ore.[191]

Hammond made no mention of gold in his speech of 1849, but the *Southern Quarterly Review* did. Mining walked hand in hand with the neglect of agriculture, to the point that people who gave up their plantations for a muddy existence barely managed to feed themselves. The editors noticed the "neglect of agricultural matters in the immediate vicinity of the gold mines, more particularly, we should say, of those in the mountain regions," where "there is little attention given to raising even the ordinary comforts of life; and the daily fare of the mining population is, in their own language, 'only just tolerable.' "[192] Here were starving men and the hardscrabble of the rush in the last place we might expect to find them. The scramble for mineral wealth in the eastern states did not, by itself, indicate the failure of rural reform, but interest in extractive industry among people at the top of South Carolina society adds further evidence that by 1850 a certain moment had passed.

Edmund Ruffin sensed its passing and descended into a depression that ended with his hideous suicide at the close of the Civil War. In 1852 he delivered a farewell address to the planters he once hoped to save, to his lifelong work, and to the Old South:

> Planters of South Carolina—I have offered to you . . . the last advice and admonition that I can expect to utter to you. . . . My burden of years, and infirmities much greater than even suited to my age, admonish me that my labours may soon close. I would deem it a reward of more value to me than will be the short remainder of my life, if you and your fellow-labourers . . . would heed my words, and fully profit by them. . . . Choose, and choose quickly! And remember, as my last warning, that your decision will be between your purchasing, at equal rates of price, either wealth and general prosperity, of value exceeding all present power of computation, or ruin, destitution, and the lowest degradation to which the country of a free and noble minded people can possibly be subjected.[193]

CONCLUSION

The Pioneers, chapter 1: "Near the centre of the State of New-York lies an extensive district of country." A great river takes its rise there; the mountains are arable to their tops; rich valleys well cultivated surround happy villages, where people value learning and worship God. The fortunate citizens live under mild laws, which inspire each to take an interest in the prosperity of a commonwealth. James Fenimore Cooper maps a world of order, tradition, and the kind of permanence that can only arise from the identification of one's own body with a native soil: "The expedients of the pioneers who first broke ground in the settlement of this country, are succeeded by the permanent improvements of the yeoman, who intends to leave his remains to moulder under the sod which he tills, or, perhaps the son, who, born in the land, piously wishes to linger around the grave of his father."

Like Cooper's Arcadia, a description of his home and the world of his father in the 1820s, restorative husbandry held together a particular landscape. Rather than glorify a young country, improvers gloried in the advantages of a maturing one and looked back on a half century since the skimmers moved on, when their towns finally grew neat and tight. Said a speaker before the New York State Agricultural Society in 1842, "At the beginning of the present century, the aspect of the country was dreary, compared with its present appearance. At that time, the arable land was worn down by constant tillage; the crops were light; the fences poor; . . . loose stones lay scattered over some of the best lands."[194] Improvement suggested a way for communities to hold or slow the extension of plowland into fields, forests, and waters, and it offered an opposite kind of change from the blaze and shift of nineteenth-century America.

Improvement flourished in a certain historical moment coinciding with economic depression in the 1820s and quickened emigration to new Wests. It represented a rearguard effort to slow the process of state formation. Its conservative quality stemmed from a belief that the United States seemed to be moving too rapidly toward unfettered democracy and unfettered expansion. Rather than direct their efforts toward state legislatures, as did the members of certain New England communities seeking the protection of communal forests and waters, improving farmers thought mostly of the natural resources in private hands. The social obligations of property, the fear of abandonment and political irrelevance, a desire for "get ahead" if not wealth, and the sense that action in the present would reverberate for generations to come inspired them to invest in a perpetual process of cultivation. The agricultural societies to which they almost invariably belonged served the purpose of generalizing individual behavior to a larger population through published papers and addresses delivered in the heat of the day at agricultural fairs.[195] The panicky intensity of restorative husbandry was the creature of a nation on the move. Reformers spoke to the fears of families and communities in the language of agronomy. In their writings and by their examples, the gritty disciplines of the barnyard preserved a culture and a landscape.

Yet this book casts improvement as a set of *practices* and not as a community activity. The object here has been the emergence of an ethic. Conservation took its rise in a practical understanding of nature. It owed nothing to purely technological advances in agriculture, which if they made farmers more affluent did so because they enabled farmers to exploit more land with less effort. Before schools of forestry came to the United States, reform-minded farmers attempted to sustain their own yields and preserve a balance in the landscape. The people at the center of this story did not reimagine their society, did not translate progress in the quality of tilth to progress in the standing of poor or enslaved people. Some in the North clung to the idea of an affluent elite, who, by persisting in the East, might neutralize the volatile acids of American democracy. Most in the South justified slavery, even to the point that they constructed a flawed land use around it, though it caused their poverty, isolation, and eventual political destruction.

Still, improvers everywhere did something remarkable. Long before the science of ecology, they came closer than anyone before them to a full (if sometimes inaccurate) sense of interdependence among organisms and of interconnectedness in nature generally. Alone among nonscientists before the 1840s, *farmers* understood that the atmosphere, the biosphere, and the lithosphere interact through the exchange of elements. *Farmers* recognized the value of landscape diversity or, in other words, that cleared spaces must be balanced by forests and waters. *Farmers* thought through what we might today call land-use economics, or the interaction of cultivation and the market aimed at the best possible adaptation of both to the environment. This speaker addressing the Board of Agriculture of the State of New York in 1821 made a striking distinction: "Agriculture, or the cultivation of the earth, strictly speaking is but an art, which teaches the best way, under particular circumstances, of tilling the earth. . . . But husbandry, or rural economy, is a science which involves the vegetable and animal economy of the whole creation, and their dependence on each other."[196] By implication, human communities joined in that interdependence.

There is no point attributing to the reformers of the early Republic ideas that they never claimed for themselves. They certainly did not invent new sciences. Instead, they should be considered progenitors of a general regard for nature that would, in the next century, become a policy of government. They provide good evidence that what ultimately became codified by professional scientists and conservationists had roots in amateur practice. At the height of their clarity, improvers articulated an ideal of progress without geographical expansion, wealth without waste, that balanced ecological production with decomposition. Yet it all changed so quickly after 1850 that we need another chapter to make an accurate accounting. Improvement simply ceased to exist in the form we have been following and became like the shards of an ancient vase strewn over a laboratory table. Chapter Three is a work of excavation carried out under the assumption that a moment had passed.

Three

Social man repays to the earth all that he reaps from her bosom, and her fruitfulness increases with the numbers of civilized beings who draw their nutriment and clothing from the stores of her abundant harvests. The fowls of the air, too, and the beasts of the field, find in the husbandman a cherishing friend. —George Perkins Marsh (1847)

I hear a voice . . . bidding Man the Cultivator advance boldly and confidently to take his proper post as lord of the animal kingdom and wielder of the elements for the satisfaction of his wants and the development of his immortal powers. —Horace Greeley (1853)

Toward Conservation

George Perkins Marsh took the floor of the House of Representatives in February 1848 to denounce the war against Mexico. He said nothing about the controversies internal to the conflict—the incompetence of politically appointed generals or the defections of volunteer soldiers. Like other intellectuals, Marsh lashed out at the contradiction of a justly constituted republic that seemed prepared to confiscate most of North America. This "useless and guilty war" not only put the United States in the company of "the most rapacious States of the Old World"; it reopened a shattering debate about the extension of slavery. Rather than simply come out in support of the Wilmot Proviso, which would have banned slaves from the usurped territories, Marsh came out against acquiring those territories in the first place. Buried under thousands of pages, hopelessly obscure among the records of the Thirtieth Congress, Marsh's remarks envisioned a very different United States from the one taking shape in Santa Fe and Monterey Bay.

A man ever conscious of his own origins, Marsh would have rolled back the expansion to the nation's first hold on the continent, calling that the only territory with moral boundaries:

> Our original limits fulfilled all the necessary conditions of national prosperity, and I much doubt whether we should not at this moment have occupied a higher place among the nations of the earth than we now enjoy, if we had been content with the inheritance our wiser fathers devised to us. We had territory, such, in proportion and configuration; that it was wholly

> invulnerable from without, and at the same time so situated as to give us the most enviable facilities for universal commerce, as well as for maritime power; we enjoyed a boundless variety of soil, climate, and natural productions; an extent of surface adequate to the sustenance of a larger population than any kingdom of Europe, and yielding the most abundant materials for industrial elaboration, the most plentiful means of commercial exchange. What more than this has earth to offer social man?[1]

How much more than what Euro-Americans first possessed on these shores do Americans require, and by what right do they claim more? The question was so completely averse to the Polk administration, so utterly lacking in political allies, that no one noticed it. Yet it was the central question for one who looked to land use as a template for tracing a nation's character. Imagine a representative from the new state of Texas rising to oppose: "Would the gentleman from Vermont please explain how a nation limited in its extent to the eastern slope of the Appalachian, with a rising population, should grow and prosper if not by taking hold of empty forests and prairies?" Prosperity can no longer depend on an outward search but must proceed by inward stages to a more perfect use, Marsh might have said. By the time he entered Congress, Marsh had begun to move beyond the familiar ethic of northern improvement into a greater regard for land.

The old project fractured. Beauty, restoration, science, fealty, permanence—remember when manure embodied them all in the form of an aromatic pile standing on the floor of a big stone barn? Farmers continued to practice the old religion, and the number of rural journals and agricultural societies increased and moved west toward mid-century. But a number of breakaway trends rising in the 1840s changed the direction and even the purpose of rural progress. The subject is best approached through individuals whose lives and ideas bridged improvement with the movements that succeeded it. Solon Robinson (Yankee, speculator, squatter, editor, critic), Andrew Jackson Downing (New York–born landscape gardener), and Marsh each began a career somewhere within the context of agricultural improvement but ended up somewhere else. The events of the 1840s and 1850s amounted to a distending of the original creed so far and so broad that no one remembered the farmers who first espoused it. Marsh has been a leitmotif in this essay because he took old-state improvement and kicked it up to a higher orbit, beyond field ecotope and farmscape to the upper levels of ecosystem hierarchy—region, hemisphere, planet. If any thinker linked the good husbandry with the conservation of Theodore Roosevelt's generation, it was Marsh.

The formal derivation of Marsh's ideas appeared in 1847, the year before his lecture in the House, when he addressed the members of the Agricultural Society of Rutland County, Vermont, on the errors of husbandry. He opened with the most tired theme of rural oratory, a stale tribute to civilization arising from agriculture. After pastoral people settled down to crop, "proper and permanent social institutions now begin to germinate." This is where most speakers trumpeted the enduring virtues of rural life and the nobility of the farm family—but not Marsh. The first lesson of settled society came hard and fast, and its meaning could not have been lost on the audience—some of the only people in the United States willing to hear and able to understand it. A people dedicated to fixed habitation necessarily regarded the world more desperately than wandering shepherds, and they understood "that man has a right to the use, not the abuse, of the products of nature; that consumption should everywhere compensate by increased production; and that it is a false economy to encroach upon a capital, the interest of which is sufficient for our lawful uses." In one sentence Marsh fused the assumptions behind northern improvement with the principles of inchoate conservation. In effect, he used the subject of agriculture as a springboard to discuss the fears he and his audience shared about the degradation of Vermont over the previous century and the possibility of its restoration. More than that, by calling nature "capital" and its produce "interest," Marsh employed a terminology that would influence conservationist thinking throughout the twentieth century. When it was tended in just the right way, nature earned and returned, and though that may have been a new idea applied to forests and watersheds, it was nothing new to farmers.[2]

The final context of these remarks is the old one. Good and orderly domestic habits—the planted orchard, the clean and seeded meadow—"tend to foster a sentiment, of which the enterprising and adventurous Yankee has in general, far too little—I mean a feeling of attachment to his home, and by a natural association, to the institutions of his native New England." Marsh's own tour of the West as far as St. Paul came back to him then, and we can know a little more of what he thought about while traveling over prairies to the edge of the Great Plains, lands he had never seen before. Years later, he felt only anger and anguish at the thought of so many sons going the same way, but to stay. Marsh tried to make the case that people who become native to their place do not leave it:

> A youth will not readily abandon the orchard he has dressed, the flowering shrubs which he has aided his sisters to rear, the fruit or shade tree planted on the day of his birth, and whose thrifty growth he has regarded with as much pride as his own increases of [stature]; and who that has been taught to gaze with admiring eye on the unrivalled landscapes unfolded from our every hill, where lake, and island, and mountain and rock, and well-tilled field, and every green wood, and purling brook, and cheerful home of man are presented at due distance and in fairest proportion, would exchange such scenes as these, for the mirey sloughs, the puny groves, the slimy streams, which alone diversify the dead uniformity of Wisconsin and Illinois!

Eastern chauvinism sounds good and righteous next to so many exaggerations of western beauty, common by the 1840s. The point is that a personal commitment to home became part of conservation as Marsh received it from an earlier generation of farmers. Farmers of Marsh's persuasion loved better what they could shape at home than what they saw in the West.[3]

Marsh said more, pointing his audience toward that greater restoration many of them already knew and practiced. Farmers determined the future of the forest. "The increasing value of timber and fuel ought to teach us, that trees are no longer what they were in our fathers' time, an encumbrance." Then he went on for pages about hydrology: "Forests serve as reservoirs and equalizers of humidity. In wet seasons, the decayed leaves and spongy soil of woodlands retain a large proportion of the falling rains, and give back the moisture in time of drought, by evaporation or through the medium of springs." The effect is to regulate the flow of water over the surface in springtime and keep the ground moist through summer. Marsh completely changed the terms of that discussion: not an encumbrance, not merely timber and fuel, not simply diversity in the landscape—trees fulfilled their most important anthropocentric function when they held humus on hillsides and maintained microclimates. This extension of principle from farm to forest, from families clearing wood on their own land to the lands that families affected by their isolated but aggregated activities, characterized a rising awareness of Nature and not simply of Land.[4]

Marsh stood on the shoulders of rural reformers to formulate his ecological ideas. No improving farmer ever explained the workings of watersheds and the consequences of deforestation as Marsh did, but they analyzed cultivation within the landscapes it affected. Here is Jesse Buel:

> In our zeal to *clear up*, we generally carry the matter to an unwarrantable extreme; every thing is cut away—the whole surface is denuded—stripped of its natural growth. We know that old forest-trees will not long bear an open exposure—that the winds will prostrate them when deprived of the protection of surrounding forests; yet the young growth, if left in clumps and belts upon the bleak borders . . . or upon portions of the farm not adapted to ploughland or to meadow, would tend ultimately to enhance its value, by the beauty which they would impart to the landscape, the shelter and protection which they would give to crops and cattle, and by the resources which they would give for fuel, fencing, and timber.

Americans had fretted about the decline of eastern forests since colonial times, so complaints of waste and fears of scarcity were nothing new in the 1840s. Yet improvers named the cause and effect and linked the balance between field and forest to the quality of cultivation. The result would be a mature landscape, only possible where people remained long enough to create it. Wood provided more than kindling and lumber, said Buel; it made a neighborhood pleasing and imparted "to old-settled districts the highest rural charms, and gives to them much of their intrinsic value." Explained Daniel Lee, editor of the *Southern Cultivator*, "If one-third of the land that has been cleared of its native forest was back in woods again, and the other two-thirds cultivated in strict obedience to well known natural laws, the rural industry of the United States would yield a quarter more than it now does."[5] A Virginia planter writing to the *Farmers' Register* expressed an ethic of restraint that became integral to conservation: "In the ten years I have been farming, I have increased the fertility of fifty acres for every one that I have injured. And in the whole ten years, I have not cleared more than was absolutely necessary for rails and fire wood; finding it easier to improve two acres than to clear one."[6]

Because of their attention to this alone, improvers rise as forebears of twentieth-century conservation. The continuum of fields and forests formed a plateau in thought, where one philosophy reached its highest elevation and the other built its base camp. As Buel put it: "To destroy, in this case, is but the labor of a day; to restore, is the work of an age." As Marsh put it: "The signs of improvement are mingled with the tokens of improvident waste, and the bald and barren hills . . . seem sad substitutes for the pleasant groves and brooks and broad meadows of [this] ancient paternal domain."[7] Had the older Buel lived into the 1840s, he might have come to the same conclusions as the younger Marsh, but he died in the midst of his work in 1839, at the age of sixty-one, just five years after founding the *Cultivator*.

In 1864 Marsh published a book that he had finished during a long sojourn in Italy on diplomatic business. He did not write about the

Old Norse culture in New England or the English language or Vermont, as he had done previously; rather, his *Man and Nature; or, Physical Geography as Modified by Human Action* presented a vision of the human occupancy of Earth. Rome had abundant physical advantages, embracing the fields of the Rhine and the Nile, the vines of Syria and Italy, the olives of Spain. Yet "vast forests have disappeared from mountain spurs and ridges; the vegetable earth accumulated beneath the trees by the decay of leaves and fallen trunks, the soil of the alpine pastures which skirted and indented the woods, and the mould of the upland fields, are washed away; . . . and harbors, once marts of an extensive commerce, are shoaled by the deposits of the rivers at whose mouths they lie." Previous generations had called these changes small and local compared with Earth's monumental workings. Marsh's innovation was to argue that these changes, placed side by side, amounted to a permanent alteration *on the scale of nature itself.*

Erosion and deforestation shaped a new world—one less habitable than what had existed before, one now well within the human realm of ethical responsibility. Marsh had learned that agriculture in particular had the potential for wholesale transformations: "Almost all the operations of rural life, as I have abundantly shown, increase the liability of the soil to erosion by water. Hence, the clearing of the valley of the Ganges by man must have much augmented the quantity of earth transported by that river to the sea, and of course have strengthened the effects . . . of thickening the crust of the earth in the Bay of Bengal. In such cases, then, human action must rank among geological influences." The old creed percolates through Marsh's dissertation, filling its most material observations with force and meaning. "The establishment of an approximately fixed ratio between the two most broadly characterized distinctions of rural surface—woodland and plough land—would involve a certain persistence of character in all the branches of industry, all the occupations and habits of life." In this way, the project harked back. In a letter to an acquaintance Marsh framed *Man and Nature* by recalling a piece of thick woodland once owned by his father "where the ground was always damp." "Well, sir, he cleared up that lot, and drained and cultivated it, and it became a good deal drier," which made it ideal for corn and grasses. Understanding environmental change on a global scale required only a farmer's sensibilities.[8] In its vigorous assertion that environments are shaped by societies, in its tactile love of place and process, the book represents the perfection of the dunghill doctrines—their transcendence of political borders.

Only rarely is there a fountainhead, as Lewis Mumford called *Man and Nature*, from which a generation of leaders and reformers come to drink. Few of the progressives who made up the cadre and constituency of progressive conservation had rural experiences, but they had all read Marsh—the man who translated a farmer's creed into language that appealed to worried urban elites. Early adherents included Charles Sprague Sargent, director of the Arnold Arboretum at Harvard University and editor of *Garden and Forest*, among the first journals to cross the line between nature as avocation and nature advocacy. Amid correspondence on garden varieties and rare conifers some found an opening for wider concerns. In a published letter dated 1889, Charles Eliot Norton of Harvard University assailed the profligate settlement of the West, calling the Oklahoma land rush a "strange, hideous, barbaric spectacle" and an example of "the prevalent misuse and waste of the natural resources of the country." He found an ally in "the late eminent George P. Marsh," who years before had demonstrated "the calamities and desolation which had fallen upon many once fertile regions of the earth through the ignorance and reckless disregard of man." Nothing could be more familiar than the tenor of Norton's remarks, yet they came out of a new corner.[9] An intellectual elite had integrated *Man and Nature* into a larger doctrine of progress without knowing what preceded it.

In a work replete with Marsh's influence, Nathaniel Southgate Shaler, a professor of geology at Harvard, wrote in *Man and the Earth* (1905) that no people have been so negligent of their soils as Americans. "Our folk developed an almost incredible carelessness in their tillage." He proposed that "the average rate of wasting should be reduced to the conditions of nature in all the areas which have been won to the plough," though he offered no specific remedies. Marsh-like, he warned that, "with rare exceptions, the fields of all countries have been made to bear crops without the least reference to the interests of future generations." The links between restorative husbandry and conservation faded in the decades that followed. Forestry became the most visible heir to Marsh's ideas.

Gifford Pinchot, who along with Theodore Roosevelt designed the most comprehensive resource policy to that time (after Roosevelt became President in 1901), also read *Man and Nature*. Years later, he reluctantly acknowledged Marsh's book: "Unquestionably it started a few people thinking," but only a few, because forestry lacked a wide readership and because "Marsh was one man, and his book one book."[10] Pinchot learned more from his privileged view growing up in the lumber business. His father, James Pinchot, made a fortune by clear-cutting the forests of Pennsylvania, and Gifford came of age in New York City—the son of a mercantile family, with the woodland experience of a monarch visiting provincial tracts. Pinchot first heard that one could study forests after his father suggested it to him in 1885, just as Gifford was leaving for Yale: "How would you like to be a forester?" In the same way that sports hunters first reacted to the decline of wildlife, the timber tycoon James Pinchot recast his best interests as the national welfare and recommended the profession of forestry to his son. At Yale, Pinchot studied with William Brewer, professor of agriculture at the Sheffield Scientific School. And after returning from graduate work at L'Ecole Nationale Forestière at Nancy in 1890, he secured a job through Frederick Law Olmsted that took him to Biltmore, the North Carolina estate of George W. Vanderbilt. His view of the United States held great potential for reform: "The American Colossus was fiercely intent on appropriating and exploiting the riches of the richest of all continents—grasping with both hands, reaping where he had not sown, wasting what he thought would last forever."[11] From Biltmore he founded the Yale School of Forestry in 1901 and from there went with his friend Roosevelt to Washington, D.C.

Pinchot in the woods represented a different line of evolution. He never knew (nor would have admitted) that his concerns and sensibilities linked back to British, French, and Dutch botanists who visited tropical islands in the eighteenth century and pronounced that colonial forests would give out under mismanagement and waste. What the historian Richard Grove terms the "Edenic island discourse" forms a parallel philosophy with improvement, with its own moral precepts derived from the Protestant theology current at the time. The observation that deforestation caused the desiccation of soils and thus aridity marked the beginning of a forestry tradition that extended through Alexander von Humboldt (the most eminent natural scientist of the early nineteenth century) to Hugh Francis Cleghorn (prescient botanist, conservator of forests in British-controlled Madras, inspector general of Indian forests in the 1850s and 1860s) to Marsh (who corresponded with Cleghorn). In short, woods defined an ethic with much of the same brooding urgency that soils inspired in farmers. Anchoring hillsides, retaining moisture, preserving temperate microenvironments, and furnishing almost the entire material culture of the world before plastic and steel—forests became associated with the persistence of civilizations. This is the project that Pinchot inherited, and little or nothing connects it to the improving farmers of the 1820s.[12]

Marsh may have exerted his greatest influence on government, and the reason has to do with the view he afforded legislators and chief executives. *Man and Nature* looked at the landscape from high above the ground. It presented a geographical panorama that government could adopt. It stressed the effects and consequences of deforestation rather than the aggregate activities that were its cause or the specific practices that would have prevented it. When Congress passed the Forest Reserve Act in 1891 and the Forest Management Act in 1897, it did not mandate practices for the better care of woods; instead, it set out to limit activities that it considered destructive to the public lands. Farmers, loggers, miners, commercial hunters: from the point of view of conservationists they removed "resources" for short-term gains, so the resources needed protection. Simply put, the burden of restraint shifted from the individual to government, or, as Pinchot put it, "The planned and orderly development and conservation of our natural resources is the first duty of the United States."[13]

Forestry alone, however, did not make up the whole of American conservation as the movement emerged in the 1890s. An agrarian branch also thrived for a time and may have reached its apex of influence with the appointment of Liberty Hyde Bailey of Cornell University to chair Roosevelt's Country Life Commission. Pinchot served as a member, as did a number of the most important names in agriculture at the time, but Bailey stood out. A scientist and editor of encyclopedias who translated the language of the laboratory into the parlance of farmers, Bailey represented the survival of high-minded husbandry into the age of conservation. He articulated the qualities of a humane (perhaps utopian) American countryside where tractors and high-intensity production always submitted to community values, enforced principles of landscape integrity, and never threatened biological diversity. Conversation that conserved not a human habitat but only the mulch for churning out the gross national product was no conservation at all to Bailey. He presaged the work of Lewis Mumford, who more than any other thinker in the twentieth century understood the fundamental need for spaces and activities at a human scale. Bailey attempted to mend the dislocations of modernism by holding fast to nurturing institutions and regenerative practices. As he wrote in 1927, "Farming is more than a means of livelihood, a farm is much more than a manufactory for the production of raw material, a farm home is more than merely a domicile. If a farmer . . . does not delight in increasing soil fertility and producing perfect specimens of his various crops, if he does not derive pleasure in watching and making things grow . . . he is in the wrong place by remaining on the farm."[14] Bailey attempted to steer national policy at the moment when population tipped toward cities and the rural constituency became the minority. He cast his work as integral to the larger project of conservation, exemplified by his service to the Roosevelt administration, but Bailey never found the audience he sought. City-born progressives had no head for land thinking and no need for it.

Agrarian conservation, with its emphasis on individual behavior, did not vanish. It appeared in the critical soil surveys published by Hugh Hammond Bennett, first chief of the United States Soil Conservation Service. Following World War II, it joined the counterculture in the form of organic farming. Scott and Helen Nearing set out to live self-sufficiently on restorative principles. Their book *Living the Good Life* (1954) tells the story of how they took over worn-down farms in Vermont and Maine. J. I. Rodale published *How to Grow Vegetables and Fruits by the Organic Method* in 1961 and spoke to household gardeners in a voice recognizable from the rural press of a century before. As Rodale defined it, organic gardening emerged from a rejection of industrialized food: "People are tired of mealy, weeks-old, warehouse-ripened produce with a tainting of insecticide or weed killer. They want quality foods for themselves and their family. Yet about the only way to insure this is to *grow them yourself.*" Wendell Berry wrote *The Unsettling of America* (1977), a jeremiad against disconnected landed practice and modernist thinking. Berry argued that American culture rises zombielike from the sterile soil of industrial agriculture. Technology and society have abandoned interdependence as a guiding principle, and more than sustenance hangs in the balance: "We can have agriculture only within nature, and culture only within agriculture. At certain critical points these systems have to conform with one another or destroy one another."[15]

A century after the founding of the *Cultivator*, at a time when New Deal conservationists confronted the desolation of rural society on the Great Plains—communities pockmarked and heat blasted by dust storms, caused by skim and scratch on a colossal scale—no one remembered the first progressive farmers. Buel's words haunt the machine pace of expansion implicated in the Dust Bowl: "We should consider our soil as we do our free institutions—*a patrimonial trust—to be handed down*, UNIMPAIRED, *to posterity*. . . . Both are more easily impaired than they are restored—both belong, in their pristine vigor and purity, as much to our children, as they do to us."[16] This is not the end of the story. Conservation and its rural legacies are only one strand among others of the disintegrating cord of improvement. The remaining strands thread across one eccentric life, or at least they can be understood that way for the purpose of brevity.

ROBINSON'S PRAIRIE

Solon Robinson appeared wherever physical and social environments collided for forty years, from 1827, when he landed in Cincinnati from Connecticut, until 1868, when he retired from the *New-York Tribune*. Backwoods settler, big-city editor, western speculator, and arbiter of taste—Robinson spent much of his life writing whatever he thought in narrow newspaper columns. He is all but forgotten today, perhaps because his work never pointed in a single direction, perhaps because he pursued agrarian prosperity with none of the nostalgia or sentimentality increasingly popular among city people.

Born in 1803, Robinson cut his teeth as an Indiana homesteader in 1831, when he promoted a city called Solon. A naturalist and church worker who visited the settlement that June described two interior spaces—that of the house and that of the proprietor: "We had our breakfast in a building which externally was quite an ordinary cabin, built and roofed with logs. Inside, however, everything was very respectable and even elegant." The library impressed the visitor, but not the atheist newspapers and anti-temperance tracts probably strewn around to goad him.[17] Robinson offered sixty-four tidy lots, close to mills and the State Road, but no one showed up. He reappeared as an innkeeper, tried the auction business, and ran a lending library in Madison, Wisconsin. Then back to Indiana, where he founded Oakland County (renamed Lake County in 1836) and again tried to blaze his name on something big, calling his neighborhood Robinson's Prairie. The land had been surveyed, but the government had not yet offered it for sale by the time he arrived, so Robinson squatted and even formed a union to protect the claims of his fellow farmers, including the occupancy rights of the Potawatomi chief Shobonier. He hardly had time to put up shelter before he was off again to tracts of prairie "entirely unsettled" after receiving a tip from government surveyors. Robinson finally came to a stop where the map went blank and put up a cabin fifteen or twenty miles beyond the last inhabitant, thirty-five miles from a sawmill, not far from the lakeshore that became Chicago.[18] He had gone to the limit, but not without his Yankee sensibilities. In the blank spaces of the public domain he hit upon ideas different from those that wilderness usually inspired.

Turning his attention to the landed economy that underlay the project of settlement, Robinson committed himself to the rapid improvement of wilderness Indiana beginning in 1837, when he wrote his first letter to the Albany *Cultivator*. As he told emigrants thinking of making the passage west in 1840: Don't come with your mahogany sideboards and gilded mirrors; come with good-quality stock, Berkshire pigs, and Durham bulls. Robinson did not discourage anyone from western settlement, but he invited easterners with caution: "Let every person disposed to emigrate, first seriously inquire whether he would better his situation or not." In 1838 he published "Proposition to Facilitate Agricultural Improvement" in the *Cultivator* and may have sought the Smithsonian bequest from Congress to establish an American Society of Agriculture in Washington, D.C.[19] Having corresponded with the *American Agriculturist* since its founding in 1842, Robinson joined the paper as assistant editor. The opportunity set off a family crisis in which Robinson finally left Indiana for New York City in November 1848 without his wife, Mariah, and their five children after Mariah refused to make the move. A year after his arrival, in January 1851, A. B. and R. L. Allen resigned as editors to devote themselves to the implement business, and Robinson renamed the paper *The Plow*. Sometime thereafter he caught the attention of Horace Greeley, who hired him to write a column for the *New-York Tribune*.

Robinson died in Jacksonville, Florida, in 1880 and is buried there, but the turning points in his life say only a little about Solon Robinson. He is best understood as a bridge between improvement and progressive farming—that amalgam of science and technology that took hold in the West beginning in the 1850s. Robinson believed that technology served the ends of social progress, that western settlement could be perfected from the outset, without a period of waste and decline. He imagined a West where farmers riding on reapers earned the affluence that came from reverent dominion over nature, not a West of thoughtless grab and exploitation. But would bald materialism demolish the fragile coalition of technology and social ideals?

Robinson and others helped to bring American agriculture to that very collision, though he never saw it coming.

Progressive farming made no fussy distinction between moral and material progress; it never wavered from the goal of larger harvests. Farmers wanted more of what they saw in the industrial economy at mid-century, and they wanted bigger yields for less labor expended. New tools offered both and manifested a transformation in agriculture. Two things conscientious farmers tended to protect and preserve were nutrients and labor. The key inventions of the 1830s blasted barriers that had limited these things for thousands of years and unleashed human labor and chemical fertility on a new scale. In the overlapping images offered by guano and the McCormick reaper it is possible to see the outlines of modern agriculture: land extensive and chemically dependent. Guano conquered the first barrier: nitrogen.

Among the looping nutrients of the biosphere, nitrogen makes the most spectacular loop. While calcium and phosphorus move from crops to animals to soils, nitrogen passes through every realm of the biosphere. As with the other nutrient elements, a lack of nitrogen is limiting to plant growth, resulting in thin stalks and yellowed leaves. Unlike the others, nitrogen is a slippery partner and easily lost in spite of its apparent abundance. The atmosphere is about 78 percent nitrogen gas and 20 percent oxygen (the rest is argon). This plenty would seem to indicate a habitable earth soaked in nitrogen, but that is not the case. The oceans are almost devoid of it, and very little makes its way into soil. In fact, although nitrogen appears in most soils, it has only two ways of getting there: lightning strikes and leguminous plants. Only energy as high as that in lightning can overcome the inert qualities of nitrogen gas and force it to combine with hydrogen to form ammonia—a plant-usable form. The other way is even more astonishing, though less dramatic to watch. Legumes absorb nitrogen from the air and convert it with special bacteria that live in nodules on their roots. For centuries farmers rotated clover or alfalfa or beans through their fields as a slow but sure way of capturing an elusive element. When guano, a rich source of nitrogen, arrived in urban markets, cultivators all over put their cash down to have it. Some had been prepared for the outlay after purchasing town dung, along with marl and gypsum, for years, but the new product delivered more than these. Suddenly there came a remarkable turn from older ways of doing things. With guano, fertility could be restored without fodder crops or cattle and without waiting for lightning to strike.

Off the coast of Peru is a series of small islands where birds have fished for anchovies in the cold, upwelling current for perhaps tens of thousands of years. The dung piling up beneath cormorants all that long time, called guano (an ancient Amerindian word), is rich in nitrogen and phosphorus. The dung stayed dry and intact in the Pacific climate, so when Alexander von Humboldt first saw the deposits they stood two hundred feet deep in some places.[20] Beginning in the 1840s, English and American firms claimed the islands for the sole purpose of selling guano as fertilizer. Solon Robinson trumpeted its benefits and even accepted a position as agent for a company that sold guano. Another firm, the American Guano Company, carried the stuff to New York and Philadelphia on ships named *White Swallow* and *Flying Eagle* and sold the gray chalky cargo for $35 to $40 a ton. Slaves and later wage laborers worked the guano mines in unimaginable conditions, nostrils burning with the acrid dust. As the supply from the Chincha Islands gave out in the 1850s, Americans looked for more. By act of Congress in 1856, and under the signature of President Franklin Pierce, individuals and corporations could stake out guano beds under legal (and naval) protection. Congress and Pierce believed that Jarvis and Baker Islands, both south of Hawai'i and within a new sphere of American territorial influence in the Pacific, constituted vital interests. The American Guano Company was organized under this act by a New York merchant and two ship captains. They called this manure "destined, by its indisputable effects, to work a revolution in the department of scientific Agriculture."[21] Consumption of guano in the United States totaled 140,000 tons in 1855.

Even supporters of guano worried about it. Robinson himself considered guano good for a jump start, but only so that farmers could continue to furnish manure themselves. He quoted a dedicated practitioner who said, "I would advise no one to rely upon guano exclusively," for while it contained more nitrogen than dung per unit, dung contained the exact proportion of the nutrients removed from farm soils. Dung embodied the cycle of nutrients in a given place; guano came from outside the loop. That may have been why even Liebig, as quoted by Robinson, warned that "a rational agriculturist, in using guano, cannot dispense with stable dung." The cost also represented something new to be considered: "Many persons object to the price of guano" at $47 a ton.[22] Horace Greeley had his doubts too. As he said about bone, another source of soil nutrients, its efficacy was not in question but "that it will ever return its cost and a decent margin of profit, is yet to be demonstrated to my satisfaction." Greeley found the stuff at $90 per ton and dismissed it: "I fully share the average farmer's partiality for barn-yard manure in preference to most, if not all, commercial fertilizers."[23] Unlike the people who would advocate the same technology in the next century, Robinson and Greeley understood the burden of additional out-of-pocket costs to farmers with little cash to spare.

Something important was happening. The purchase of this substance, not fashioned on the farm but imported from far away, became (in the minds of some) an emblem of progressive farming. Guano stands as the first chemical input, for although it came from an island and not a laboratory, although no one manufactured it, it filled the same niche and arrived in the same way that synthetic chemicals would soon arrive: in a bag from a supply company. A correspondent from Durham, England, to the New York State Agricultural Society understood the enormous implications of guano and its similarity to the first generation of industrial fertilizers and pesticides: it prepared farmers for what was to come, "prepared them more willingly and more generally to receive those preparations and mixtures which Chemistry will by-and-by be able to compound, with the view of meeting the wants of every soil and every crop."[24] The last bit was no afterthought. Whereas convertible husbandry reflected every local quality of the environment, guano paid no attention to the world around it and worked the same way wherever it landed.

Guano may have been a quiet addition to many farms, but it thumped a shock wave through the ecology that unifies producers and consumers. If animal manure completed a circle on the farm, even one complicated by the increasing amount of produce going to market in the nineteenth century, guano represented a breaking and straightening of that line. With guano and all that followed it, the biological foundation of American society became a one-way transfer of material from some point of extraction or production to the farm where it went into the crops, and from there to consumers. Nothing came back. Robinson had no intention of replacing barnyard manure with commercial fertilizers, but the higher concentrations in guano could not be ignored. Robinson claimed that a single wagon could carry enough guano to dress twelve or fifteen acres, while the same wagon filled with stable dung would dress an acre or less to produce the same results.[25] Higher concentrations meant greater cost efficiency in transportation and labor. And if farmers wanted a source of nutrients that did not ask them to dedicate land to grass and fodder, that did not require them to milk the cows and clean the barn, guano suggested a different kind of farm. With this, more than with any other modern addition, cultivation began to change. It began to depend on institutions beyond the furrow—not simply for markets or transportation but for the continuation of the biological process itself.

The second barrier to larger harvests that farmers had long confronted was more formidable than nutrient levels. Labor had only rarely been a subject of the improvers. Northerners near Boston, Philadelphia, and New York City knew where to find hired hands. The northern farm evolved under the burden of labor scarcity, in a land where almost anyone could own a farm. The southern plantation never existed before slaves or indentured laborers. No plantation crop—not tobacco, rice, indigo, cane sugar, or cotton—could be harvested by machine in the nineteenth century. Regardless of where they lived, reformers assumed that the methods they advocated could be integrated into the existing labor system. Edmund Ruffin, among others, strained to argue that slavery did not hinder but enhanced improvement. The shape of any form of production will reflect the quality and the quantity of labor (in addition to land and capital). Planters wasted land in order to conserve labor, and the tiny fields of New England provide good evidence of scarce human power. Farmers and planters owned a certain amount of labor, so the crucial question was how to make the best use of it. Anything that offered to magnify that power would change the landscape and agriculture.

Implements magnified labor. In virtually every instance, they allowed people to work greater areas while expending less time and energy per acre or bushel. A boy with one horse could rake a wider path in a shorter time with one of the new "claw-toothed" cultivators than he could have without it. One invention completely reconfigured the relationship between land and labor in North America by introducing the idea of mechanized harvesting. The McCormick reaper consisted of a low-to-the-ground blade that moved back and forth when driven by gears attached to the wheels. As the horse moved forward, a rotating reel pinned the wheat against the cutting surface and laid it down on a catchall. Cyrus McCormick first tested his design in 1831, and Obed Hussey came up with another in 1833. McCormick received his patent in 1834 but confessed that the most famous machine of its kind had little "practical value" until his second patent in 1845. Even after that, it clogged up and broke down in wet stands. Anything but wheat in a perfectly upright position, dry and free of mud, might be trampled into the tilth rather than collected.[26]

It is easy to forget that for most of the last ten thousand years people gathered wheat while bent over or on both knees; one hand grabbed a clump of stalks, and the other cut them with a blade. The wheat had to be thrashed (or separated from the stalk) and winnowed (the kernel removed from the bran) by hand. Yet it says something for the tenacity of old ways that implements caused little discussion, even among the most urbane farmers, until the 1830s. Thomas Fessenden, editor of the *New England Farmer*, listed a number of inventions in his book of 1835, saying that though they had proliferated over the previous decade, they were ignored by farmers, who "very absurdly retain their old implements, though convinced of their inferiority." There is no mention of the reaper in the index of the first volume of the *Cultivator* in 1834. In his 1843 report *Improvements in Agriculture and the Arts*, the commissioner of patents only once mentioned machines or implements in connection with improvement; instead, he named geological surveys, agricultural societies, rural periodicals, and sobriety as pillars of improvement. One reason for the long neglect of these tools is obvious. Gadgets with names like the Albany seed-drill, Clinton's improved corn-sheller, Grant's fanning mill, and Mott's portable agricultural furnace, in addition to plows and patented reapers, did not necessarily deliver the results their makers promised. That and the cash outlay necessary to buy them (before banks mortgaged equipment) priced most implements beyond the reach of the majority.[27]

Farmers had other reasons to criticize the reaper, though only a few of them did. These footdraggers—some of the only people to question what would soon become the new regime—saw the reaper for what it was: an expansionary machine that did nothing to improve the condition of soil. It only put more land under till. Said Daniel Lee in 1848: "Improved plows, cultivators, and other implements, have been placed in the hands of millions of industrious laborers, to scratch, skin, and bleed the virgin soil for a few years, which, with the assistance of a bright, burning sun, and washing rains, soon consummates a very satisfactory degree of general desolation." Then he asserted a familiar alternative: "Great industry and mechanical skill, in consuming and wasting the elements of bread and meat, which a kind Providence has placed on and near the surface of the earth, are more praised than they deserve. . . . If we can contrive to leave our children a reasonable surface of good farming land, we need not be at the trouble of converting its soil into current gold for them." Nathan Cook Meeker, who helped to invent the irrigated colony as a founder of Greeley, Colorado, observed that the machine only tied farmers more tightly to the market with little gain: "Agricultural machinery does not work for the farmer at all; it only increases production, and the same proportion passes into other hands, by the jugglery of commerce, as when grain was cut with a sickle, and the farmer must see that he totally fails to get as his own any part of this increase." Meeker wondered "whether there is much of any profit in the use of the great line of agricultural machinery—that is, to the farmer himself."[28]

The machine had arrived, and quite regardless of those who equivocated, it mowed a new path to rural progress. "When the census of 1860 is published in full," wrote the commissioner of agriculture in his first annual report to Congress, "the inexorable logic of its statistics will astonish the world, and prove to every intelligent mind that agriculture is the grand element of our progress in wealth, stability, and power." He referred to the seemingly inexorable rise of implements, whose total value in 1860—not counting those manufactured privately on farms—came to $17,802,514, an increase of 160 percent in ten years. A writer in the same report announced that "in nothing is the advancement of modern agriculture more conspicuous than in the rapid improvement of the tools and machinery of the farm," and then he detailed the state-of-the-art fifty-acre homestead, circa 1862, including three plows and a subsoil plow, a corn planter and seed drill, one combined mower and reaper, assorted scythes, manure forks, and hand hoes, and two wagons. It all came to almost $1,000 in equipment. Farmers would seek to distribute these costs by turning out their reapers on fields equal to their new labor capacity. An observer close to the action, Henry Howe, understood the reaper. With four thousand of them assembled each year at McCormick's factory in Chicago, "mainly for the use of the Western farmers," the reaper stimulated production and accelerated the settlement of western North America. Even more strongly, Howe compared it to another famous machine: "McCormick's Reaper may be said to be to the great West what Whitney's cotton gin is to the South—a machine of incalculable advantage in developing the resources of the country."[29] It is remarkable—and not much appreciated—that implements redefined the leading practices of cultivation, endorsing the same land waste that improvers condemned in the 1820s, calling that "progressive farming."

Looked at another way, guano and the reaper stand like bookends on either side of a new soil economy. They may appear to indicate different paths to higher production. The first suggested greater investment on existing land, or intensive cultivation. The second used land extensively, as farmers always had; it simply caused a jump in scale.[30] And yet the two also implied each other. Where the reaper increased the number of acres a farmer could harvest, it made fertilizing land from within the resources of a farm less likely if not impossible. Expanding the scale of cultivation to take full advantage of the reaper suggested one of two things. Either farmers would continue to depend on fresh land without a thought of restoration, or they would have to find some source other than the eighty head of cattle necessary to keep a quarter section continuously manured. In this same line, it is not surprising that Solon Robinson found guano more common on Virginia plantations than in Connecticut. Robert W. Carter of Sabine Hall on the Rappahannock had two thousand acres in cultivation and boasted that he would never again cultivate wheat without guano. He mixed it in at two hundred pounds per acre and yielded (a paltry) eight bushels, though another field yielded fifteen. Carter called guano the savior of lower Virginia. Planters recognized that this product enabled them to plant more land than they could possibly fertilize from within their resources.[31] So it would be everywhere when agriculture entered the machine age. The larger areas for cultivation made possible by the reaper created an opening for commercial fertilizer because any machine that magnified labor and expanded the scale of production did so without regard to the capacity of a farm to recycle its fertility.

On his visits to plantations and farms, Robinson repeatedly noticed the way things looked, the design of buildings, and the attractiveness or ugliness of the rural scene. The aesthetic prejudice for order and ornament inherent in improvement also sprang into its own career.

Only a short distance separated field and farmhouse. Improvers of the 1820s saw them as manifestations of the same economy, with the house as the center of combustion for the domestic product. Every example of food and fuel passed through the kitchen, literally through the fire. Yet at just the time when Hammond and Ruffin worried about the persistence of the planter class in South Carolina, a striking transition took place near the northern cities. The distance between field and farmhouse widened. Improvement had always been a philosophy of home, defined as region, state, county, or farm. It had also been the philosophy of those with the ways and means to embellish. A proper home from the architect's sketchbook had certain conventions: seated in a lawn, garden gated, Gothic pointed, shrubs to frame the view. The affluent urban class that consumed the latest landscape painting and breathed deep the wilderness sublimity in high-above-the-clouds mountain hotels wanted some of that romantic repose at home. On carriage rides just beyond city limits they saw the raw material for a different kind of countryside and a different conception of improvement.[32]

According to Robinson, they often saw the downed fences, broken storm windows, and shingle-worn existence of "Farmer Slack." Farmers throughout New England and New York fell into hard times as wheat cultivation moved west. The commissioner of patents estimated in 1849 that of the twelve million acres of improved land in New York, "eight million acres are in the hands of three hundred thousand persons, who still adhere to the colonial practice of extracting from the virgin soil all it will yield." Harvests had fallen from near thirty bushels an acre to eight in seventy-five years. Making that trend seem even worse, crops grown on rocky soils in small fields no longer paid off when they met crops from the table flats of Illinois in major markets. Those who attempted to make up for the decline of wheat by selling dairy goods to New York City often lacked the skill and efficiency to carry on a more complicated business. In the last of his many incarnations, Robinson came to the realization that farms just outside the cities, which should have been some of the best kept, looked like some of the worst.[33]

Robinson espoused all the practices of convertible husbandry, yet after he returned to the seaboard, he emphasized their connection to domestic economy and an aesthetic of rural living. It seemed to him, as it had to a generation of improvers going back to George Washington, that the American countryside too often looked like wretched neglect. Disgusted during a visit to Connecticut in 1849, he wrote,

"Much of the land . . . is covered with bushes, or miserable little half-starved patches of cultivation, or with shanties that are a degree, at least, below the western log cabin. And this is within the sound of the City-Hall bell. And this is 'the age of agricultural improvement,' is it?" On the railroad to New Haven he despaired of "the same old stone walls and rickety rail fences, bush pastures, bog meadows, alder swamps, stony fields, and scanty, because unmanured, crops, that were to be seen in the same places fifty years ago." Robinson is clearly part of the older tradition but with this addition: he moved neatness and appearance to the forefront of his concern and mixed old-time rural economy with something new—a degree of polish beyond the means of regular farmers. "A country house without a lawn! it is a house in a desert! It is not a structure in the midst of beauty. There is nothing—not even expensive statuary, flowers, and shrubbery—that adds so much to the surrounding embellishments of a farm-house or suburban residence as green grass upon a well-kept lawn, and it is a beauty that is permanent and inexpensive. . . . It tells, too, of art and industry in man, since lawns are seldom, if ever, found in a natural state."[34]

While hunting for fine and foul landscape in Massachusetts, he praised an exemplary farm for its thrift and elegance. In fact, the owner did not live there most of the year; he operated a hotel in New Orleans and only summered in New England. Robinson described the seat as unimaginably delightful amid "a broad, smooth, grassy lawn, beautifully ornamented with a great variety of trees and shrubs, with ornamental cuts for flowers in the sod, the whole forming a lovely, shaded retreat." The owner knew how to move forest trees from the woods to exactly where they would enhance a vista or launch a bucolic mood. More for the eye: a flower garden behind a rustic fence; a picturesque cottage for servants. Robinson could have chosen any prosperous farm to paint his "contrast" to slumping Connecticut but chose this one, where he noticed only highbrow ornament. There is no mention of the farm at all on a property of 120 acres—the home and its amenities absorbed all of his attention.[35] Robinson, who for fifty years espoused hard-boiled agronomic advice, also believed that a countryside ought to look good—and he was not alone in that view.

Other people wanted the farm without the slack. Improvement had imported strong aesthetic and class preferences along with convertible husbandry, and the influence of John Sinclair and Arthur Young goaded the members of elite agricultural societies to new degrees of ornament. Those preferences often brought up the rear behind the more tactile needs of barnyard economy, and they could be mocked, as they were by John Lorain, as frivolous and even hindering to the full realization of an American middle landscape. An emphasis on strict economy and restoration in the context of arable husbandry obscures the rising interest in ornament that characterized the period after 1840, especially among the members of the Philadelphia and Massachusetts Societies for Promoting Agriculture. Horticulture, racehorses, full-blooded stock—different kinds of breeding became popular for pleasure and status, and their popularity among the affluent signified that agricultural experimentation had gone out of fashion as a public-spirited act. The lawn opens another view to the subject, since it resembles something with real utility (pasture or young wheat) but typically had no practical purpose. In the suburban borderlands outside New York City and Boston during the 1840s and 1850s, aesthetic desires began to cleave from restoration to become a force all their own.[36]

What, exactly, is a lawn? It may be any swath of sod for grazing. Its earliest definition has it as an uncultivated patch, an open space between woods. Yet it eventually shaped a green plane that unified foreground and background, holding a landscape together like a frame. When Capability Brown recommended turf as part of the design for English country houses in the eighteenth century, the lawn became beauty in the midst of structure. This is where it remained for American designers, many of whom understood its connection with pasture yet cheered the triumph of loveliness over utility. A key example is Lewis F. Allen's *Rural Architecture* (1853), clearly part of improvement literature, but the new domestic variety. Designs for ice- and smokehouses, barns and chicken coops, are functional yet poised.

The houses are a mix and match of American styles, including a plantation mansion, along with designs for parks, woodlands, and cottages for "the skirts of estates." Allen misunderstood the utilitarian aesthetic of many rural people when he asked, "Why should a farmer, because he is a farmer, only occupy an uncouth, outlandish house, any more than a professional man, a merchant, or a mechanic? Is it because he himself is so uncouth and outlandish in his thoughts and manners, that he deserves no better?" Farmwork can get a little messy; it can even be "rough, laborious, and untidy." Yet Allen saw no reason why rural living could not display "neatness, order, and even elegance and refinement within doors." The effect was to redirect the domicile away from the barnyard in a manner that threatened the whole homestead with too much "order." The lawn formed an important part of this design.

We are no longer in Lorain's Pennsylvania or Hammond's South Carolina, and yet it would be a mistake to come away thinking that this new form of improvement shirked hard work and utility. Consider Allen's favorite exterior decoration, ornamental livestock:

> The man who has a fine lawn, of any extent, about his house, or a park adjoining, should have something to graze it—for he cannot afford to let it lie idle; nor is it worth while, even if he can afford it, to be mowing the grass in it every fortnight. . . . This ground must, of course, be pastured. Now, will he go and get a parcel of mean scrubs of cattle, or sheep, to graze it, surrounding his very door, and disgracing him by their vulgar, plebeian looks, and yielding him no return, in either milk, beef, mutton, or wool? Of course not.[37]

None among Allen's audience could afford to let good pasture go uneaten, and they expected to turn a profit on the usual products. So the livestock were working ornaments, and that makes them more interesting. In fact, this literature persisted in the same reform of landscape, was enshrined in the same state and local agricultural associations, and expressed the same desire for permanence as what had come before. It simply pressed a more individualistic emphasis.

Self-definition and conspicuous consumption trumped feigning republican simplicity.

The people within view of city lights wanted romantic farms—real-world counterparts to the landscape painting they loved. Said one beautifier from New York: "A good farm should not be characterized alone by its trim fence and its straight furrows, but it should also be an object of beauty . . . a scene of surpassing loveliness." Famous artists might create loveliness by depicting the delicate glade and the brilliant meadow, but "shall the claims to taste and genius be denied to a farmer who can create such a scene, and be awarded solely to him who can transfer it to the living canvass?" "These higher accomplishments," continued the speaker, "constitute no part of practical farming, because [they] contribute nothing directly to a farmer's prosperity." An earlier generation might have left it at that by affirming that the desire for beauty must always be subordinate to the economy of plain farming, but disciplined pragmatism had softened by 1841. Even God would fail to qualify as a practical farmer, for God has covered the earth with objects "whose surpassing beauty is their only utility."[38] Under these ideas, the eastern suburbs were becoming a cultivated world unto themselves, free from the burdens of field husbandry. Amid concerns such as the proper appearance of cattle standing near the front door, how long would the barnyard hold out? With its pools of urine, rotting vegetables for the hogs, and piles of feces, it might nauseate rather inspire. The people who would soon explore the Sheep Meadow in Central Park wanted the simulacra of rural life. They found their "apostle of taste" in the dean of American landscape design—the Martha Stewart of his time. He turns out to be another complicated thinker in the distended history of agricultural progress.

Andrew Jackson Downing designed cottages for middle-class and affluent people who wanted to retire in comfort to the countryside. His signature design sits prim and picturesque on a lawn that may or may not have been attached to a farm. It is a farmhouse in gesture at least, though it looks too prissy to handle a working tilth. The dark and dramatic-looking man who dropped out of school to work in a nursery providing ornamental and orchard plants to the rich might be easy to dismiss as the wizard of the soft and emotional suburban house, but his writings and the context of his designs demand more attention.

The surprising thing about Downing's *Rural Essays*, a collection published after his early death in 1852 at the age of thirty-seven, is the fierce argument in favor of restorative agriculture apart from any aesthetic judgment. In a number of short works, Downing emerges as a conservationist who attached the condition of soils to the stability of the societies they nurtured. In "The National Ignorance of the Agricultural Interest," written in September 1851, he compared Americans to "a large and increasing family, running over and devouring a great estate to which they have fallen heirs, with little or no care to preserve or maintain it, rather than a wise and prudent one, seeking to maintain that estate in its best and most productive condition."[39] Downing knew the look and feel of prosperity in landscape, and he also had a sense of its fragility:

> Now, there are two undeniable facts at present staring us Americans in the face—amid all this prosperity: the first is, that the productive power of nearly all the land in the United States, which has been ten years in cultivation, is fearfully lessening every season, from the desolating effects of a ruinous system of husbandry; and the second is, that in consequence of this, the rural population of the older States is either at a stand-still, or it is falling off, or it increases very slowly in proportion to the population of those cities and towns largely engaged in commercial pursuits.

Mocking the shortsighted state legislature, he asked their question, "What matters it . . . if the lands of the Atlantic States are worn out by bad farming? Is not the GREAT WEST the granary of the world?" Canals and railroads wrote a one-way ticket for fertility, he said; they "bring from the west millions of bushels of grain, and send not one fertilizing atom back to restore the land. And in this way we shall by-and-by make the fertile prairies as barren as some of the worn out farms of Virginia."[40] Not that any of this forms a contradiction to his love of picturesque landscape. The investment in real estate that a

Downing home represented, and the extensive landscaping and lawns necessary to frame it just right, would lead any owner to cultivate permanence.

Downing suggested his own solution to the problem of the eastern states, one that fit nicely with his overriding goal of infusing the New York countryside with beauty: orchards. Fruit trees produce crops of great value from soil of lower-than-average fertility. "If it is too difficult and expensive to renovate an old soil that is worn out, or bring up a new one naturally poor . . . in the teeth of western grain prices, [one] may well afford to do so for the larger profit derived from orchard and garden culture." Not only did the orchard give the rural East a new way to calculate natural advantages over the West, but graceful stands suggested a more refined rural life—and one that paid. Small, picturesque farms gave "agreeable residence for a gentleman retiring into the country." Small and irregular fields, curving paths leading to the stables and the barn, orchards in a parklike setting surrounded by lawns, and the house and other buildings "of a simple, though picturesque and accordant character"—all would lend "an air of novelty and interest about the whole residence." An orchard made any farm into an arbor and attracted the gentlemanly prestige of horticulture. One local from Westchester County, just outside New York City, reported that fruit growing had already "begun to change the aspect of many farms which were formally devoted to ordinary productions, such as stock, grain, and the grasses; and they are now becoming more profitable as orchards and vineyards."[41] Never had economic survival looked so comely and smelled so refreshing.

If the image of the English countryside haunted improvement from the start, providing an example with such authority that it virtually became part of the genetic makeup of reformers in the North, Downing brought the same aesthetic to people looking for rural repose. The orchard married all the class pretensions of horticulture with the hard economic concerns of farming. As improvement, the orchard stood for polish and permanence. But Downing failed to see that the cottage culture he promoted would eventually supplant a functioning countryside. A people losing touch with yard, tilth, and woodlot only wanted to gaze at forest trees; they made a hobby of identifying plants they did not know by daily use and sought emotive experiences in contrived landscapes. The farmhouse became the home in the garden, and soon enough fields and orchards disappeared altogether from the suburban borderland. Agriculture moved farther away from the lives of many people. It moved west. A few easterners stubbornly committed to the idea that human perfection could be achieved by a more improved form of settlement bypassed the old Northwest and the Great Plains after the Civil War.

Looking back less than fifty years on the settlement of western Georgia and the states of the Southwest, the editors of *Soil of the South* had nothing good to say. Whites from the eastern states had only recently taken possession of lands where the Cherokee and Choctaw had planted and hunted for centuries. One removal ordered another. The evicted Indians walked a thousand bloody miles to "Indian Territory" on the dry-land plains, a place utterly different from what they had known. The editors of 1851 came down hard on the Anglo-American occupation of Alabama, Mississippi, and the other states of the Old Southwest. This fresh and verdant territory had run down the same old gullies: "We are reaching a crisis, where the basis upon which all this great and glorious structure rests, is about to perish. The soil of the South . . . is being rapidly impoverished." A pernicious cycle finally threatened to engulf the entire South: "Our old lands are worn out, our new ones are going. Gullies and red hill, and pines and broom sedge, tell now, where once was many a fertile plain." People will depart, they said, "until we have no longer a West to which they can retreat." It all went down the ditch so quickly that the settlement of the Southwest seems more akin to vandalism than to agriculture: "Men not yet old, will note in the brief history of their own recollection, how the wilderness has fallen, and the gullies and barren hillsides have multiplied, under the constant wear and tear of a system of culture which has abstracted all, and returned nothing to the earth."[42]

Eugene Hilgard saw the same decay. A German-born soil scientist, Hilgard had a sharp sense that arable land lay vulnerable and only half realized without considerable intellectual and financial investment. The state of Mississippi appropriated neither during his career as state geologist from the 1850s to the 1870s. The only purpose of a state geologist, from the point of view of governors and legislators, was to boot up and tramp the country looking for mineral deposits shimmering at the bottoms of riverbeds. Hilgard winced and refused to cooperate with small-minded boosters "who imagine that all the meaning and object of a Geological Survey is the discovery of mines." In an unadorned report of 1858 he scorned the extractive assumptions of those who put him in charge. Hilgard confronted not so much the leadership of Mississippi as its citizens, many of whom insisted that all was well as long as "lands are not yet exhausted" and they could continue to pilfer. He wished all the pilferers a long, hard journey and *auf Wiedersehen!* "Their departure . . . will scarcely be felt as a loss to the community they desert."[43]

No one listened. Still at his post during Reconstruction, Hilgard saw the aftermath of the Civil War as an opportunity to reconstruct land use. He spoke to cotton farmers about the "public calamity" that had unfolded over less than half a century, hitting them hard, bringing into question the human wreckage that made way for some of the very people sitting before him. "Well might the Chickasaws and Choctaws question the moral right of the act by which their beautiful, park-like hunting grounds were turned over to another race, on the plea that they did not put them to the uses for which the Creator intended them. . . . Under their system these lands would have lasted forever; under ours, as heretofore practiced, in less than a century more the State would be reduced to the condition of the Roman Campagna." Hilgard sang the same song of destruction and conservation and confronted the same assumptions as those of the previous generation in the old states. In an echo of William Gilmore Simms, the South Carolina writer who lamented emigration, Hilgard reflected not on fresh destinations but on the homeplace and the improvements that made life dear: "Refined, cheerful and pleasant homes are everywhere the marks of an intelligent, refined and moral community."[44]

All of this made way for Hilgard's own emigration—not the same passage that brought planters to Mississippi but a transition from improvement to something else. His diatribe before the cotton farmers seemed to indicate a course for some other place, the Far West. Early in 1874 he left for California, where he founded the first university experiment station in the United States and began to propound the idea that agricultural affluence would sprout from irrigation and the scientific planting of soils. The other side of the continent offered more than just fresh starts; it lacked people with ingrained ways of doing things. Another figure who thought in much the same way at the time was also the nation's most visible editor. Horace Greeley had a complicated relationship with agriculture and the West, and, like Hilgard's, his projects represented the emigration of ideas to where they could be fully realized.

Horace Greeley knew how to farm, though he had left his family's homestead at the age of fifteen to work (like Jesse Buel) as a printer's apprentice. Decades later he bought fifty-four acres in Chappaqua, New York, where a murmuring brook soothed his "burning, throbbing brain." There he set himself to experiment with the biological process at the foundation of all political economy. Greeley's laboratory yielded *What I Know about Farming* (1871)—not a tome for practical farmers but rather a series of essays about what Greeley "firmly believed." The core of the book lies squarely within the tradition of improvement ("The fairest simple test of good farming is the increasing productiveness of the soil"). Yet the one thing Greeley most wanted farmers to know moved him beyond restorative husbandry: land must receive the most intensive possible cultivation or else farmers would be left behind as the industrial economy burgeoned. "I do firmly believe that the time is at hand when nearly all the food of cattle will, in our Eastern and Middle States, be cut and fed to them." Cattle allowed to "roam at will over hill and dale" browsed all the succulent plants and left the weedy ones to propagate. Pasturing appeared to underutilize land.[45] Greeley's sense of balance and tightness in landscape ran right down the line with Downing and Marsh. As he told Indiana farmers years earlier, in 1853: "If all the labor now devoted to farming, throughout the Union, were wisely concentrated on one-half the land, our annual product would be much larger, our lands would appear far more productive and valuable."[46]

The new industrialist commanded fuel, fire, and human energy, so the new farmer needed to become "wielder of the elements." One thing captivated Greeley for twenty years and fortified his conviction that agriculture would not lag behind in a world full steam ahead with railroads and telegraphs. On the short-grass plains of Texas in 1871, standing in the desert at the heart of the continent, Greeley imagined a waving stand of green timothy: "In the great Future which Science and Human Energy are preparing, Artesian wells, bored to depths of a thousand to fifteen hundred feet, will be sunk on every arid plain, and near the head of every capacious valley wherein water is deficient." This was a matter not simply of better living through applied water but of the ascent of civilization from barbarism to "commanding altitude" through the control of nature. Greeley had already seen the future during his first western tour, back in 1859, when he visited a community of New York farmers who had removed to Utah in 1847 seeking haven from violent religious persecution. Determined to keep land and community together in an arid environment, members of the Church of Jesus Christ of Latter-day Saints (the Mormons), under the leadership of Brigham Young, invented a system of irrigation and social coordination that deeply impressed Greeley.[47]

It was the West that impressed Greeley. The *New-York Tribune* covered the region ceaselessly and called itself "The Great Farmers' Paper." Greeley even featured a special edition for castaway subscribers in California, Oregon, and the Sandwich Islands. He shared a firm commitment to the development of the region with Solon Robinson, the western/agricultural columnist for the *Tribune*. The two editors might have been separated at birth: New England–born with a rough-hewn aspect and a knockabout quality, feverish writers and critics, believers in social progress, and lovers of the big tour. Both believed that first settlers did not need to pass through an era of waste and decline but could jump ahead. Lacking entrenched authority and with a population needing instruction in the ways of a new environment, the West made the ideal laboratory. After Robinson's retirement in 1868 Greeley hired Nathan Cook Meeker, and the dream of a progressive rural community suddenly took form. Meeker was a war correspondent when he came to Greeley's attention. He had been an itinerant poet, a schoolteacher, and a devotee of Charles Fourier, the French socialist whose writings inspired in Meeker a desire to found a utopian cooperative. As his son recalled, the family knew Mormons who told them about the Rocky Mountains, and Meeker informed his children that they would one day leave Ohio and "go to the Mountains and found a community of a few families, removed from the noise and frivolities of society."[48] A year or so after Meeker's arrival in New York City, Greeley is reported to have said to him: "I understand you have a notion to start a colony to go to Colorado. . . . I wish you would take hold of it, for I think it will be a great success, and if I could, I would go myself."

A colony. The earliest examples of western irrigated settlement expressed one necessity and innumerable desires. The necessity had to do with agriculture in an arid climate. Isolated individuals could never divert enough water from permanent and seasonal streams to sustain crops like wheat and corn on a commercial scale. Only a community willing to share the costs of irrigation and the hardships of isolation would survive the first year. Such a settlement should not take the shape of the rectilinear townships of the East, where people spread out more or less evenly over the surface of the earth in a pattern that reflected the even distribution of rain. The colony would have to be small and tight and would need to be established in relation to a specific watershed. John Wesley Powell, an anthropologist and government explorer who navigated the Colorado River in the 1870s, came to this conclusion, and, like Greeley, he had visited the Mormons first. As for the desires, a community supported by a sure watercourse could form an autarky and carry out any philosophy without resistance, like Puritans in the New England wilderness.

Meeker founded the Union Colony, including the town of Greeley, north of Denver, on principles of temperance and social equality. But this was no government-run kibbutz, and settlers needed money: at least ten of the settlers had to come with $10,000, twenty with $5,000, and on down to $200. The poor would eventually be invited to join. The colony took the form of a village modeled after Northampton, Massachusetts, with farm lots of forty to eighty acres adjoining town lots. Barely a year after its first residents moved in, the hamlet of Greeley earned this description from a tourist's guidebook: "Two hundred houses already, and not a solitary one last spring. The inhabitants all have more or less capital, and so they will escape the poverty-stricken childhood of most pioneer settlements."

The founder turned out to be long on idealism and short on practical knowledge. He declared that irrigation cost little more than fencing, the only other similar investment that an eastern farmer could imagine. The initial estimate for a complete system came in at $20,000; the true cost was $225,000. Novices expected too much from the combination of land and river flow. David Boyd, author of an early history of the colony, reported that settlers arrived convinced that "irrigated land never wears out." Greeley himself predicted forty bushels of wheat per acre at the new colony, but for the first two years the average fell below fifteen. Other things went very wrong. The cooperative mission degenerated into squabbles about costs and the design of irrigation facilities. All the fruit trees died. Then Indians captured Meeker with his wife and daughter, apparently in the aftermath of a recent massacre in which they probably played no part. The Indians killed Meeker and released the women after twenty-three days.[49]

The Union Colony became one of those points where the electricity building up between East and West arced. This project, perhaps more than any other at mid-century, bridged eastern notions of moral and material progress with western possibilities. Yet just as the colonists tried to acclimatize certain crops and trees to new conditions of temperature and soil, they themselves changed. Two visitors make the point. The colonists received Solon Robinson like an ancient sage when he stopped there in 1871 . . . until he opened his mouth. He afflicted them with a two-hour address, most of it "made up of buncombe, spread eagle platitudes about the climate, scenery and the immense capabilities of the soil." The inhabitants knew better what it meant to make a living in a foreign environment; they needed engineering, not limp attempts at inspiration. Fifteen years later, in 1886, they welcomed the help of a young man named Elwood Mead, who served briefly as Colorado's assistant state engineer and came to the Union Colony to measure the carrying capacity of a critical ditch. As director of the Bureau of Reclamation in the 1920s, Mead presided over the era of giant-scale water storage and personified the conservationist/planner/engineer. Yet throughout his career he encouraged government aid to small settlers, believing that individuals and private land companies had failed to "create the kind of rural life this nation needs." Mead summoned government to finance the cost of irrigated settlement and combat the waste and misery of haphazard schemes with a sound alternative, and, in this way, to adopt a socially progressive land policy in the arid region "as a public matter." Cooperative community, established on intensive agricultural principles, would become the project of the various states, each with its own rural land commissions, like the one Mead headed in California during and after World War I. But Mead's communitarian vision died with him. The social ideals that once clung to irrigation, and agriculture itself, no longer compelled policy makers or the urban majority after 1919. Though the improvement of land would continue to describe acres freshly cleared and under till, its more enlightened meaning was deserted.[50]

Like the head on an ancient stalk of wheat, improvement shattered in the 1850s and its seeds fell every which way. It grew into progressive agriculture and landscape design and even played a part in the formation of at least one socialist community in the West. Pockets of the old creed persisted, as in the diversified farm established by Ellwood Cooper in Santa Barbara, California. Cooper was born in Lancaster County, Pennsylvania, in 1829. Once in California he spoke against the first generation of chemical insecticides and in favor of planting one-quarter of all farmland in eucalyptus trees. His ethic of landscape conservation owed something to Marsh, especially his belief that the world's climate had changed due to deforestation. He took words directly from *Man and Nature* when he said worriedly, "The earth is fast becoming an unfit home for its noblest inhabitants." Cooper's love of green countrysides links him back to the rural Pennsylvania of his youth.[51] As the distant ancestor of organic farming, the old husbandry of the 1820s might be upheld as evidence that history took a wrong turn in the 1850s, that things could have been otherwise than the technological morass that furnishes food for the developed countries in the twenty-first century. The truth is more complicated and less satisfying. Historians know well that every era and epoch is a transition from something to something else. Improvement cannot be cherished as some saintly relic—perfect in death—toward which seekers after environmental stability can direct their prayers. Simply consider that the builders of industrial agriculture also believed in the permanence of settlement and efficiency in the use of resources. They had no doubt that breakthroughs in the invention of new plants, new machines, and new fertilizers represented the future security and affluence of their society. They shared important similarities and a common creed of progress with nineteenth-century agrarian reformers.

STRIVING AFTER HARMONY

Fertilizers, pesticides, reapers, and harvesters constituted vaulting achievements in the ability of farmers to make food and fiber from sunlight and soils, but they stood for the abandonment of an idea. At the end of the nineteenth century the modern farm no longer looped its nutrients, and that has had all sorts of implications for the environment and society of the United States and the world. The association of technological progress with environmental degradation and the virtually uncontested transition from improvement to progressive farming force some difficult questions: Was there ever a stable agriculture in North America after Europeans arrived? Was improvement a fundamental innovation in land use, a missed opportunity, or simply a stop along the way to industrial agriculture?[52] In a sense this book has meditated on these questions. Yes, an ecologically stable agriculture existed in North America after Europeans arrived, but few people used it to reconfigure land use, and it held the vital center of public attention for only a few decades. What does it say about improvement that its triumphs faded so quickly? How do we account for the fact that if we pull back the focus, improvement seems to fit into a much larger continuum of technological progress? To what extent should improvement be implicated in its own demise?

Improvement rests firmly on the continuum of progress that historians thread through the Renaissance to the Industrial Revolution. It was fully the heir of the marriage of science and capitalism that took place between the sixteenth and eighteenth centuries in Europe and North America. The organisms whose combination composed convertible husbandry formed a biological lattice that extended the ability of people to exercise their will in the natural world. The complex should thus be considered a technology even if it did not rely on anything mechanical. But what kind of technology? Is all technology the same? Lewis Mumford lights a path through the history of invention and social change with two useful words: *polytechnic* and *monotechnic*. Polytechnics derive from an ancient lineage and combine functions and skills shared by many people in a society. Polytechnic tools can be used in different ways and allow people to develop skills and exercise control over their work. In a polytechnic society no single method of doing anything dominates, and no authority dictates technology. Think of craftsmen with their various guilds employed to build a cathedral. Mumford insists that medieval ways of doing things were not backward or slow: "Until the seventeenth century, this polytechnic tradition performed the feat of transmitting the major technical heritage derived from the past, while introducing many fresh mechanical or chemical improvements," including inventions as radical as the spinning wheel and the printing press.

Monotechnics are defined by the concentration of authority in the proliferation of technology: "time-keeping, space-measuring, account-keeping, thus translating concrete objects and complex events into abstract quantities." The projects of monotechnical thinking proceeded from different assumptions. Take mining in Europe from the sixteenth to eighteenth centuries. Mumford writes that it depended on a disregard for human life akin to that of warfare, that it caused environmental destruction and poverty wherever it appeared, and that it offered the possibility of profit far beyond human needs. The definitive technology of the nineteenth century, the steam engine, first came into use to pump water from the bottom of mines. Worst of all, the entrepreneurs and inventors who have sponsored monotechnic thought over the last two centuries "sacrificed human autonomy and variety to a system of centralized control that becomes increasingly more automatic and compulsive."[53] Monotechnic inventions can also be decentralizing (think of the automobile), but they always dictate the way a thing is to be used. All of these describe technology divorced from and opposed to the emotions, desires, playfulness, and spontaneity that people strive for in their lives.

The question is, which frame of mind did convertible husbandry represent? Certainly it did not represent centralized authority, but could it have led to that? Agriculture escaped the grip of monotechnic transformation for a very long time because the basic breakthroughs of the eighteenth century aimed at harnessing steam and waterpower for manufacturing, both highly capitalized processes with no immediate application to farming. This neglect accounts for the "backwardness" of agriculture into the twentieth century. Yet it is inescapable that improvement stood at the beginning of the very technological surge that eventually brought farmers to adopt DDT and tomatoes enhanced with flounder genes.

Let's go back to brass tacks: convertible husbandry was not mechanized technology, so it did not dictate a uniform process, frozen in flywheels and gears, unalterable by any individual. It stemmed from an established principle (cycling nutrients); it was brought about by familiar elements (fodder crops and animals); all of which came together in methods requiring new applications of human and animal labor (shifting cultivation, feeding and soiling of cattle in pens). The elements and methods could be adjusted to fit any situation, and competing ideas suggested by local people in their associations substantially changed the way farmers applied the overall principle. It seems crucial that no one could simply "plug in" convertible husbandry without asking which crops and animals functioned best in a given environment; or in other words, the whole system could not be automatically reproduced in any two locations. A sustainable system does not deplete its bank of fertility and incorporates positive feedback loops that can be initiated by people with little technology. The elements composing convertible husbandry lay like paint on a palette that could be configured for restoration and permanence along a spectrum of held capital, from Mount Vernon to the backwoods of Pennsylvania. Finally, it took a long time for farming to incorporate the divisions of labor definitive of factory production. Until the middle of the twentieth century, people engaged in rural work performed each and every task, and nothing required specialized skill. All of this argues for improvement as the apex of polytechnic cultivation.

Turning the question over: the methods of improvement fit together like the parts of a machine, like the gears of a clock. Plants and animals could not be standardized or dictated, but the breeding of plants and animals became popular, resulting in species that agricultural societies recommended as the most profitable. When the goal became higher production and more money, domesticates soon reflected these desires. State-sponsored societies, and later experiment stations and land-grant universities filling the same niche with a more explicitly scientific mission, advocated certain combinations.[54] This looks like the alliance between government and technology that Mumford so dreaded, though in its most inchoate form. The incomparable example of this fusion had antecedents in the era we have been considering. Solon Robinson first proposed a National Society of Agriculture in 1839 to disseminate information from an authoritative center. The organization never took hold and folded. Twenty-three years later, in 1862, Congress established the United States Department of Agriculture. The USDA did not simply propagate improved methods—it became the Church of Information and Technology (with its own missionaries) for millions of modernizing farmers. Its experts eventually embraced any machine or chemical that promised increased production regardless of how technological change would affect farm families or the environment. Robinson intended none of this, but his advocacy of the collected methods of convertible husbandry—increasingly based on expert experiment and propagated by organizations with expanding authority—carried a forward momentum that did not stop with guano. These methods form the recent ancestry of the most monotechnical of institutions.

Improvement flourished in the moment between the rise of a capitalist and an industrial society, a brief time in which material progress coincided with an ecologically benign form of agriculture.[55] For a moment, leading farmers advocated a philosophy in which rising affluence might function within the economy of nature. For a moment, the most advanced practices rectified farming with ecological production and decomposition. Conservation emerged from the same moment and embraced the tension of the controlled disturbance at the core of civilization. Conservation did not challenge basic assumptions of material progress; it *recast* progress as timber left standing, as waters running clear, as habitat undiminished. While conservation proved to be a durable policy of government, improvement fell off the tightrope it walked over technological exploitation. It fell into neglect. The good husbandry was good because it triumphed in its place and paid its debt to soil, but it failed historically. The few decades of its preeminence tell of a fundamental instability, not in the dunghill doctrines, but in an unsettled people who did not know their borders.

Epilogue: Fredericksburg

> What hath been done may be done again. Old Arts when they have been long lost, are sometimes recovered again and pass for new Inventions.
>
> —Jared Eliot, *Second Essay on Field Husbandry* (1748)

Yellow mustard flowers rise like scattered fire towers above a canopy of green oats. Dogs dart around underneath, sniffing for rabbits, as David Kline reaches each crowning plant and pulls it out by the roots. I am walking beside him at eight in the evening, the taste of sweet grass in my mouth, my hands skimming the belly-high forest as we move along. In the distance, across a row of cherry trees, is the farm belonging to one of David's sons, where we cut hay the day before. A woodlot is the only border between oats and sky. We call out to each other whenever we spot mustard and take turns sauntering over to get it, though David finds more than I do. This work is always done by hand and on time, or the weeds will go to seed and make more work next year. We lay our bundles in the private road that connects the various parts of these eighty acres to the homesite on the county road and an additional forty on the other side. Opening a box that stands at the corner of the oats, David says, "I want to show you the bluebirds' nest," and uncovers four violet eggs. There are woodpeckers, bobolinks, and two kinds of swallows. Woodchucks live under the cherry trees, along with owls, foxes, possums, skunks, snakes, quail, mice, deer, and too many other birds to list. Gesturing toward a few rows of sweet corn, he tells me, "It's for us, but I have to have enough for the raccoons—they love it."

David and Elsie Kline do not farm like other people. They never view wildlife as a threat to their livelihood. They use no pesticides, herbicides, or genetically modified organisms. They never spray anything. Feed corn will be harvested, dried, and stored in cribs for winter fodder, while the oats will be eaten off the stem when thirty-five dairy cows chew them down and manure the field at the same time. The Klines' implements for mowing, baling, binding, reaping, cultivating, and threshing date from the 1930s to the 1960s, and all their field traction is provided by draft horses. They accept no government support and say they need none. They produce nothing for sale on the Chicago Board of Trade; says David, "Let those speculators speculate on themselves; they aren't using me as their pawn." Whenever possible they plant their own seed instead of buying it. By choice, they have no electric power in their home and own no automobiles. They buy few consumer goods, distrust centralized authority, and dress plainly. The Klines are Amish farmers. They are not dead-end holdouts left over from the agrarian nineteenth century; rather, their system of food production represents the future of American agriculture.

Interstate 71 crosses state highway 83 near the town of Wooster, an hour south of Cleveland. From there it is only a few miles to a region defined by the villages of Fredericksburg, Mount Hope, and Berlin. I saw the last of American roadside culture for three days when I left the commercial strip in Wooster, with its dominating Wal-Mart, for a county road. Something about a broad and aged opening in the northern temperate forest brings out a longing in me that I cannot fully account for, makes me feel that I will never catch up to it, never know it. Things here fit together in common pieces; farmhouse, white barn, cement silo, black buggy, Jersey cows, and the crops interrupted by woods in all shapes and angles to the horizon. The hilly topography opens brief vistas over tiny valleys, revealing colonies of farms. Settlement is surprisingly dense, a sign of the intensity families attain in cultivation. (Standing in the oat field, I counted fifteen silos.) I was ready to ditch my car, when I pulled over to ask directions from a young woman in a dress and bonnet. She told me I had arrived at David Kline's farm. I met Elsie in the driveway, and she told me they were baling hay. A moment later I was sitting on the floor of a long wagon, bumping and lurching away from the county road.

In the few days I spent with the Klines, I saw how their cultivation anchors their material lives. It is not true that the Amish use no motorized machinery. They simply refuse to use it for fieldwork where it would transform production from a biological to a machine scale. Farmers operating with the latest technology can mow and bale in one operation while sitting alone, fully enclosed behind air-conditioned windows. The Klines need every available family member to pull in the timothy grass (seven people that day, eight counting me), and they like it that way. Elsie and I sat down with the dog and talked for a long hour, waiting for David to return from a lower field. When one of their daughters appeared with milkshakes a while later it all made sense. The work of the farm is the life of the family, so it must always embrace rest and pleasure. It makes no difference that tractors haul things back and forth or that the Klines switch on a diesel generator to create suction for milking. The principle behind the policy is not difficult to discern. The Amish hold these values above all others: anything that undermines their ability to cohere as a community of neighbors and linked families, anything that isolates them in their work or places production for profit ahead of the collective process, is prohibited. David adds a corollary: no practice will be allowed to denigrate the wholeness of land and its capacity to sustain wild as well as domesticated animals.

Agribusiness managers and their advocates at the United States Department of Agriculture represent the counterfactual point of view: nothing could be more irrelevant than Amish farming. Nice and good that some people still do things the old way and manage to support themselves. The Klines and their neighbors make central Ohio pleasant to drive through, and their tenacious rejection of popular culture creates a fascinating enclave. Generalize Amish ways to the world's agricultural production, however, and a billion people would be dead in a year. The world has no more high-quality farmland, but it does have a rising population, so all the high-tech tools devised to feed people—especially those much-maligned genetically modified organisms—have become the saviors of humankind. We have created a world in which they are absolutely necessary. The rising costs and increasing complexity of food production have caused the rapid depopulation of countrysides from Argentina to India, contributing to the explosive growth of cities. All those people, who once fed themselves, now need to be fed by larger and larger concentrations of government, capital, and technology. Like it or not, so the argument goes, the high-modernist farm is the only option for the soon-to-be world of ten billion.

Yet just because something is necessary in the present does not mean it will be around for very long, nor does it mean that it is fostering of human and nonhuman communities. If agriculture has taken up just a minute in the day of human history, industrial agriculture has been a nanosecond. No one has any reason to believe that it will survive in its present form. It tends to destroy the very systems it depends on by polluting and overfertilizing lakes and rivers, by causing soil erosion, by radically simplifying biological diversity, and by requiring the constant combustion of fossil fuels. Industrial farming is more complicated than this, of course, and a technological gradient extends from the factory farms of the Imperial Valley in California to the fields of Holmes County, Ohio. Some of David Kline's non-Amish neighbors use the assortment of products that defines them as "tractor farmers."

Yet the vulnerabilities are almost identical in every case, and it remains to be seen whether the petrochemical-genetic complex will find stability or whether it possesses all the fortitude against disturbance of a Roman arch—its perfect tension failing with any loose stone. An insect or blight destructive to a key species of rice or wheat might evolve that will not die by any known pesticide or genetic modification. Crude oil might soon run out or become too expensive, driving up the price of food. The fundamental natural body sustaining settled society—topsoil—has disappeared in regions all over the United States, replaced by fertilizers that require manufacturing and transportation. If any of these systems were to fail, a billion people would be dead in a year. The more likely scenario is that family farmers who attempt high-modern agriculture will continue to fail. If that does not narrow the world's food supply, it narrows the limits of human freedom. When none derive their own subsistence or provide food for small communities, the countryside will belong to great capital. As Wendell Berry once put it, the question of the survival of the family farm is really the question of "who will own the people."

American consumers have had no reason to protest, since industrial agriculture has always served them first and foremost. The single accomplishment of industrial agriculture should not be overlooked: there is an astonishing amount of food. Hunger has not been eliminated from the United States, but its causes, at least right now, cannot be linked to high commodity prices. The handful of companies that provide chemical and biological products make abundance their stated goal, and they deliver it. Yet abundance has an entirely different meaning to small farmers, most of whom purchase the same fertilizers and seeds as factory farmers but without the same capital to back them. Rigging up the entire process from fertilizer to harvesting, paying the big suppliers every step of the way, exposes small farmers to every sudden drop in price, every dry year, every rise in interest rates. Think of them as people standing neck-deep in water; they can drown under any rise in the river. The response of agricultural economists has been, "Get big or get out," but that is not a prescription for good food, a diversified landscape, or the reign of community values over the countryside.

David Kline's manure-centered husbandry, not much different from what American reformers and their British tutors urged and implemented back in the 1820s, represents an alternative—a progressive occupancy of land for the twenty-first century. No matter how unlikely the prospect that people the world over will take up small farms like they once did (first of all, governments would have to encourage it), that is no reason to reject the Amish as unfit for the future. On the contrary, there could be no land management better suited for a small and crowded planet. Amish farming is highly productive and environmentally stable and represents a profitable way for families to remain in control of rural places. David Kline's land thinking is traditional without being nostalgic, practical without nodding to technology. And while industrial agriculture still has its viability to prove, the Amish hold fast to practices that are four hundred years old. Amish farming is not modern, but it might be postmodern.

I found David Kline in the timothy stubble, and we talked for a few minutes before going to work. He wears the distinctive beard-without-a-mustache and plain buttoned pants and suspenders. His features are soft and tanned under a straw hat, and he speaks with only a hint of the German or "Dutch" that is the Amish vernacular. He was born in 1945 on the farm he now owns, received his education in a one-room schoolhouse, and had never left home before the Vietnam War, when, as a pacifist objector, he was drafted into service at a Cleveland hospital. He worked briefly in a garage-door factory before he and Elsie took over the homeplace. The Kline farm was first deeded in 1825 by a farmer who bankrupted the soil after three generations. David's grandfather bought the property all ragged and dumped on at a sheriff's auction in 1918.

Once the hay has been raked into lines, the baling begins. Elsie steers the team while we stand on the wagon behind the sputtering machine, which scoops up the grass and binds it into thirty-five-pound blocks. We take turns stacking the bales as they come through the chute, slipping and staggering back and forth like two people dancing on a slab of ice. The straw stabs at my soft hands and covers my forearms with green dust. Then I travel on the heavy cart to the barn belonging to David's son, Tim, where I throw the bales onto a conveyor belt so that David and Tim can load them neatly into the loft. It was two in the afternoon when I started. We did not finish until eight or nine, and the family had been working since sunup.

I am interested in the profitability of Amish agriculture but tread gingerly over the subject at first. The margins between profit and loss among farmers can be wrenchingly narrow. Millions of dollars can pass through a family's hands, though they may end up keeping only hundreds. The Klines do better than that. Dropping his voice and looking at me square, David intones, "It's extremely profitable." Each year he grosses $2,000 per cow, compared with the $200 to $300 profit common on industrial farms.[1] His harvests may not be quite as large as those on farms where Monsanto has determined the exact combination of crops and chemicals, but they're "very close." When he grew wheat, David harvested seventy-five bushels per acre. For comparison, I found two reports from different parts of the country. Michigan wheat farmers collected a record-setting sixty-seven bushels per acre in 1999—an accomplishment that local people attributed to "prayers and technology." Kansas farmers brought in an average yield of only forty-six bushels per acre in 2000 under good conditions.[2] Just as important as its quantity, David's wheat cost him almost nothing to grow—just two bushels of seed per acre, selected from his own reserves (though first purchased out of a seed catalog and representing modern breeding). His principle tool for reaping and binding is a simple mechanism with a rotating reel manufactured by the McCormick-Deering company. Sitting in the shed next to an equally vintage double plow, the reaper looks like the first one ever made.

No one in the county is making a fortune this way. Mean household income for Fredericksburg, Ohio, in 1989 was $23,750, but income from farm self-employment was $30,000—the highest category in the county. The estimated market value of land and buildings per farm in Holmes came to $349,203 in 1997, compared with $449,748 for the United States as a whole, but that $100,000 difference is not at all what it seems. The average American farm enclosed 487 acres in 1997, yielding a per-acre value of $923. Farms in Holmes are much smaller—just 122 acres, for a per-acre value of $2,862.[3] The people maintain this value and a high quality of life even though they dedicate a significant portion of land to feed five thousand horses every year. Yet for all this, income, no matter how high or low it may be, is not the best indicator of Amish success, because they have eliminated the need to shell out cash for all sorts of things. They own no large machinery requiring monthly payments and purchase no chemical preparations for seed or soil. They pay little or nothing for insurance, fuel, or child care. They grow at least a portion of their own food and give food to each other. Most of the cherries I picked went to family and neighbors and the rest to market. Farms pass to sons whenever possible, preventing (or internalizing) the most significant source of debt in any rural society. And there is never any need to hire labor for wages no matter how large the task. Community provides the only real insurance they have, and it carries no price.

The most indispensable economies come from wily agronomy. The Amish strategy for competing with big-business dairy farmers is to radically lower the cost of feed by planting high-quality ryegrass or clover with an energy and protein profile to rival the traditional combination of alfalfa and corn.[4] Allowing cattle to graze appears to violate the core principle of convertible husbandry. The dung is not collected. In fact, nineteenth-century farmers often planted fine grasses and then "penned" their animals in the field, letting them manure the ground where they stood. For the Amish, that method saves all the labor of harvesting the hay and spreading the manure. The method requires far less corn than usual methods of feeding, opening land for other purposes; and because forage grass is perennial, it grows whenever weather permits, whereas corn has a specific season. Most conventional dairies yield five to seven thousand pounds of milk per acre, but skilled graziers have been known to produce eight to ten thousand. Cash savings due to this modern adaptation of an ancient shortcut protect the Amish like a storm door against low milk prices. Tim, who recently purchased a farm of eighty acres, will pay off a $200,000 mortgage in ten years without difficulty. "Some people are getting $12,000-a-month milk checks," Tim tells me, while the cost of a typical operation is not more than $1,000.

This is not the way farmers are supposed to talk. The televised image of farmers selling out, no longer able to continue because of mounting debt, bitter and weeping at the sight of their combines and kitchen tables at auction, has been common in news and documentaries since the 1980s. As any farmer from the hard-hit upper Midwest might explain, the reasons have nothing to do with the ability of rural people to raise commodities. They have to do with the fatal collision between prices and debt. All through the 1970s farmers enjoyed high prices for their crops and low-cost financing for land and new equipment. Thinking the good times would never end, many went deeply into debt in order to purchase the newest machines. Some bought combines costing up to $60,000 and took out additional mortgages for additional acres. Boom expectations inspired these purchases more often than any felt need. As Kathryn Marie Dudley writes in *Debt and Dispossession*, "A newer, bigger tractor is not just a more efficient way to plow your field. It is also a sign that you are keeping up with the times. . . . New technology, in this sense, is not so much the cause of 'progress,' but a way of *representing* it."[5] When prices slipped and the prime rate bounced from 6.8 percent in 1976 to 18.9 percent in 1981, families owing more money on their loans than the total value of their equity lost it all. Every newspaper story seemed to carry the same photograph of men with hands in their pockets looking down at the ground, their giant machines stuck and silent.

The worst farm crisis since the end of World War I never hit Amish farmers. The people of Wayne and Holmes suffer the defections of young people to town life, and the total number of Amish farms has declined since 1987, but not because the "plain people" fell into the technology trap. Government representatives have told farmers for a century that they need to secure themselves against the hard and unpredictable winds of the market economy by constantly increasing production. It is ironic, then, that Amish farmers, who never took that advice, live more resiliently (in this respect) than the great majority. Amish farmers are some of the best farmers in the world and the preservers of a genuine land ethic. The cycle of nutrients that recreates land is not an antiquated idea; it expresses a fundamental ecological principle that has maintained Amish livelihood on this continent for two hundred years. Manure and grazing figure in the curricula of agricultural colleges and are regarded as cutting-edge ideas in environmental thought. There is no mystery about these methods, and David advocates them every chance he gets, convinced that others can take them up. Bucking every trend of contemporary American society, he says, as though calling out over the hills, "We need more people on the land!"

In the morning Elsie and one of their daughters talk about larkspur recently seen in the strawberry patch, but a fawn that may be hidden in one of the timothy fields, where the baler would kill it, is a more pressing subject. Never did I see or hear any attempt to control or eliminate varmints, though summer is the peak of the growing season. There is a sense among the family that wild things enliven domestic spaces and that domestic spaces do not fracture but create animal habitat. David goes still further: "By working and farming the way the Amish traditionally have done, we make our place more attractive to wildlife. Should we be removed from the land and our farm turned into a 'wildlife area,' I'm almost positive that the numbers and species of wildlife would dwindle."[6]

This is the message of *Great Possessions*, a series of essays in the form of a journal that David published in 1990. Throughout these essays it is nearly impossible to separate the tasks of the farm from the family's love and longing for creatures. Once while David was mowing along the edge of the hayfield, a female bobolink flew up from the ground before the cutter bar. "Quickly stopping the team, I soon located the nest with its five newly hatched young. Using the cut hay, I built a flimsy canopy over the nest with the hopes of saving it." Another time, the Klines stepped all over each other, rattled the buckets, and startled the cows after a younger son came running into the barn yelling, "Geese!" There may be no better proof of the philosophy than David's warm regard for fencerows—scruffy and overgrown borders that serve as sanctuaries, where uninvited guests and fellow travelers thrive. In the fencerows lives an animal almost universally detested by cultivators. Woodchucks eat from the hay, the garden, and the young corn. Farmers shoot them by the thousands every year, but David champions them, seeing them as an indicator species for the farm's capacity for wildness: "As long as there are fencerows, there will be woodchucks."[7] Land is a total concept, and agriculture is just one of its uses. Failure to account for woodchucks and bobolinks takes farmland out of its context, cuts it off from life, isolates it in its work.

A fitting counterpart to *Great Possessions*—its long-lost cousin—is *The Wooing of Earth* (1980) by René Dubos, a French microbiologist who, after helping to discover antibiotics, criticized the eradication of disease organisms. Dubos may have been the ultimate ecologist. By looking at nature from the point of view of microbes, he recognized a continuum between the inside and the outside of the body. He conceived of human health as an *ecology*, claiming that the species inhabiting us typically exist in benign tension with our immunity. So whenever public health officials wield antibiotics to wipe out pathogens in a human community, they simply hang out a vacancy sign for others to fill the same niches. Humans and disease organisms have a long history together, with widespread sickness occurring only under certain conditions, such as poverty, and not simply because of the presence of pathogens.[8]

Dubos looked at the human occupation of the earth the same way. The two have a long history together, with destruction occurring only under certain conditions. He never ignores proliferating extinction and pollution. The book is certainly not the work of a Pollyanna. Instead, Dubos sees a deeper concord in the aching love between humans and the earth they have created. In a manner totally averse to orthodox environmentalism, which tends to regard people as disease organisms, he argues that nature needs humankind in order to develop its better qualities: "There is no doubt that people spoiled the water economy and impoverished the land when they destroyed the forests of the Mediterranean world. But it is true also that deforestation allowed the landscape to express certain of its potentialities that had remained hidden under the dense vegetation. . . . Ecology becomes a more complex but far more interesting science when human aspirations are regarded as an integral part of the landscape."[9] The same unbridled destruction of trees and soils ripped up the ground between eastern Pennsylvania and western Illinois during the 1820s, including what is now David Kline's Ohio. Yet centuries later, who would trade Nottinghamshire, the Po valley, or Holmes County for what had been there before? The worst disruptions of the past have given us landscapes "more diversified and emotionally richer" than the wilderness they replaced and just as deserving of preservation. The humanized earth consists of places deeply familiar, profoundly home, and all the more beautiful for their distance from first nature. When David Kline asserts that wildlife flourishes on his farm to a degree that it would not if the forest came back, Dubos echoes him: "We can improve on nature to the extent that we can identify these unexpressed potentialities and can make them come to life by modifying environments, thus increasing the diversity of the Earth and making it a more desirable place for human life."

Some might take this as a glossy version of the same attitude that caused the global crisis in the first place. There is an acrid smell to the notion that rising concentrations of greenhouse gases represent yet another example of humans making a home for themselves on Earth. No one who knows the value of the Brazilian forest would advocate its conversion into rural landscape. David Kline and René Dubos roundly reject shortsighted projects and short-term gain. They also reject industrial capitalism as the defining environmental relationship of our time, with its bottomless appetites and grinding externalities. They forward the idea that good management is a basic human desire and that the world does not furnish evidence of a fatal mismatch between *Homo sapiens* and nature. Amish environmentalism holds that people should be like the woolly fencerows, standing stalwart for the demarcation of cultivated spaces while keeping the borders open.

The sun is setting by the time we finish weeding the oat field. The buggy has been ready all day, so we ride for the pleasure of it as evening comes on. Taking the reins I feel a groove near the shoulder of the blacktop, which I first noticed on my way in, where the hooves of Amish horses have pounded out a place for themselves in the road. We turn around at the next intersection and pause for a moment in the gravel drive between the barn and the side of the house, while the horse sweats and breathes. I sit with David in his study the next morning. They get up to milk the cows around four o'clock, then take an hour or two of rest between six and eight o'clock so as not to make the workday too long. We talk for more than my allotted hour, causing me to rush our farewells as two men, one dressed the Amish way, one wearing a baseball cap, arrive to conduct some business. I follow that worn-in place near the shoulder, imagining how it might be used to map out Amish country, but it's gone by the time I get to Wooster.

NOTES

ONE

1. Brady, *Nature and Properties of Soils*, 6–9.
2. Spafford [Agricola, pseud.], *Hints to Emigrants, on the Choice of Lands*, 5–8.
3. What any given plant requires of each element varies widely. Macroelements, or those required in relatively large quantities, are carbon, hydrogen, oxygen, nitrogen, potassium, calcium, magnesium, phosphorus, and sulfur.
4. Lord, *To Hold This Soil*, 45; Brady, *Nature and Properties of Soils*, 431–36, 455–57.
5. Adams, *Heartland of Cities*, xvii, 54–55, 165. For an interpretation that stresses rising salinity, see Jacobsen, *Salinity and Irrigation Agriculture in Antiquity*, 11–12.
6. Danhof, *Change in Agriculture*, 252. Crop productivity had fallen off in many eastern regions by the second decade of the nineteenth century. Upland farmers in New York, for example, had expanded production by clearing new fields but finally ran out of space. Emigration followed. Taylor, *William Cooper's Town*, 387; Spafford, *Gazetteer of the State of New York*; Jones, "Creative Disruptions in American Agriculture, 1620–1820"; Carman, "Jesse Buel, Albany County Agriculturalist," 244.
7. Vaux, *Address Delivered before the Philadelphia Society for Promoting Agriculture*, 8; Nicholson, *Farmer's Assistant*, 174; Stilwell, "Migration from Vermont," 70–71, 184–86, 232–33. Atack and Bateman consider the degradation of soil to have been an important cause of emigration from the Northeast in *To Their Own Soil*, 77.
8. For more on the meaning of progress and internal improvements connected to the building of the Erie Canal, see Shariff, *Artificial River*.
9. The term "Agricultural Revolution" is beset with contradictions among historians. Kerridge's argument in *The Agricultural Revolution* is the most convincing—that a technological takeoff began in England in the 1560s, when a policy of sound money and secure private property allowed creative peo-

ple "to use their native wit and inventiveness in seeking their own profit." The earthquake in production began in nonmechanical soil practices and draining. Duncan has written in *The Centrality of Agriculture* that the spread of an ecologically benign agriculture among the landed classes during the eighteenth and nineteenth centuries in Britain was both capitalist and modern, without bold advancements in mechanization and before industrialization. Clearly, the earlier innovations prefigured implements. Also see Chambers and Mingay, *Agricultural Revolution, 1750–1880*, and Kerridge, *Farmers of Old England*. McClelland's *Sowing Modernity* overemphasizes the importance of implements and does not consider advancements in pasturage, fertilization, or rotation part of a revolutionary shift.

10. *Cultivator* 5 (March 1838).
11. *American Farmer*, 2nd ser., 1 (1839).
12. Another source estimated a circulation for the *Cultivator* of 20,000 and the same for the *Genesee Farmer*. A list of 132 names might not seem like a very convincing sample, but without the work of Sally McMurry there would be no sample at all. Few of the members of the county agricultural society appeared on the list, suggesting the limits of those organizations as indicators of the popularity of improvement. McMurry, "Who Read the Agricultural Journals?" The numbers quoted by Jesse Buel come from Demaree, *American Agricultural Press, 1819–1860*, and Gates, *Farmer's Age*, 341.
13. They are known through the work of Avery O. Craven, whose book *Soil Exhaustion as a Factor in the Agricultural History of Virginia and Maryland, 1606–1860* (1926) first made the connection between destructive agriculture and frontier settlement. For a recent estimation of Craven's work, see Kirby, *Poquosin*.
14. Tull, *Horse-hoeing husbandry*; Darwin, *Phytologia*; Eliot, *Essays upon Field Husbandry in New England*; Beer, *Inquiry into Physiocracy*; Cato, *Cato the Censor on Farming*, 14.
15. Barger and Landsberg, *American Agriculture*, 6, quoted in Atack and Bateman, *To Their Own Soil*, 264–65. Merchants and bankers—not farmers or planters—became the first industrial financiers because they knew both the supply of raw material and the location of demand and because far greater sums passed through their hands. Porter and Livesay, *Merchants and Manufacturers*; Danhof, *Change in Agriculture*, 104.
16. "The Speculator," *Carlisle Republican* (June 8, 1819). I am informed here by a number of important works belonging to the "social" interpretation of the early national countryside, especially Henretta, "Families and Farms," 15. Clark navigates between the "social" and "market" schools in a synthesis that gives the market a rich social context in *The Roots of Rural Capitalism*. Also see Merril, "Cash Is Good to Eat."

17. Kulikoff, *Agrarian Origins of American Capitalism*, 34–37, 82–90.
18. *Plough Boy* 1 (June 5, 1819). According to the editor, the paper's name was meant to evoke the Green Mountain Boys of Vermont who fought in the Revolution.
19. Appleby, "'Agrarian Myth' in the Early Republic," 268.
20. Kulikoff, *Agrarian Origins of American Capitalism*, 218; Henretta, "Families and Farms"; Colman, *Address Delivered October 2, 1839*, 11.
21. Strickland undertook his tour at the request of the British Board of Agriculture to gather information about land and labor in America. He titled his report *Observations on the Agriculture of the United States of America*. Also see Herndon, "Agriculture in America in the 1790s."
22. Strickland, *Observations*, 4.
23. Ibid., 22, 26, 31, 48, 49. For an alternative view of America and a direct response to Strickland, see Tatham, *Communications concerning the Agriculture and Commerce of the United States of America*, 48. Tatham, however, did not contradict Strickland's essential criticism and said of the Virginians, "It is found hard to leave the beaten path of a custom which has long been skimming the cream of nature, and is eternally on the search for *more new land*." Tatham's emphasis.
24. Bordley, *Summary View of the Courses of Crops in the Husbandry of England and Maryland*, 11.
25. Washington, *Letters*, 6, 12.
26. Ibid., 11–12; Jeffreys [Agricola, pseud.], *Series of Essays on Agriculture & Rural Affairs*, 5, 12.
27. Garnett, "Defects in Agriculture," 54. Sixteen years later, in 1834, Edmund Ruffin proposed his *Farmers' Register* in a broadside: "The low state of both the practice and profits of agriculture in Virginia, is admitted and deplored by all. . . . [O]ur lands remain impoverished." Edmund Ruffin, Proposal to Publish *The Farmers' Register*, Virginia, 1834 broadside page, in Brock Collection, Ruffin Family Papers, Correspondence and Documents, 1811–1892, box 80 (61), Huntington Library, San Marino, Calif.
28. George Washington to Arthur Young, Dec. 5, 1791, in Washington, *Letters*, 22. For the origin of these ideas, see H. J. Habakkuk, *American and British Technology in the Nineteenth Century: The Search for Labour-Saving Inventions* (Cambridge: Cambridge University Press, 1962).
29. Howitt, *Selections from Letters*, 199–200.
30. Darby, *Tour from the City of New York*, 15–19.
31. Ibid. The literature is filled with examples of this criticism. Consider this from the *Carlisle Republican* (July 6, 1819): "When you plough, don't break up more land than you can enrich. It is a fault among us, that we improve too much tillage land, and don't make it rich enough."
32. "Mr. Madison's Address," *American Farmer* 1 (Aug. 20, 27, Sept. 3, 1819).

Also printed as Madison, *Address Delivered before the Agricultural Society of Albemarle*, 162.

33. McCoy, *Elusive Republic*, 150–60.
34. *Mr. Clay's Speech upon the Tariff*, 10; Pearson, *Notes during a journey*, 63.
35. Samuel Hopkins in an address delivered at Batavia on October 18, 1820, quoted in Mathew Carey, *Address to the Farmers of the United States* (1821), 8–13. Eighteen sheriff's sales took place in one Pennsylvania county in March 1819, including twelve farms and a number of mills, according to *Niles' Register*, reported in *Carlisle Republican* (Jan. 26 and March 23, 1819); *Niles' Register* (Aug. 7, Sept. 4, Oct. 23, 1819); Flint, *Letters from America*, 248, quoted in Rezneck, "Depression of 1819–1822, a Social History," 31.
36. *Niles' Register* (Oct. 31, 1818; June 12, 1819; and July 28, 1827); Dangerfield, *Era of Good Feelings*. Sea island cotton sold for even higher—up to 65 and 60 cents a pound.
37. Rothbard, *Panic of 1819*, 11–13.
38. Taylor, "Anniversary Address Delivered before the State Agricultural Society of South Carolina . . . 1843," 203.
39. Feller, *Jacksonian Promise*, 40–43; Cayton, *Frontier Republic*, 125–26.
40. Coxe, *An Addition, of December 1818, to the Memoir, of February and August 1817, on the subject of Cotton Culture, the Cotton Commerce, and the Cotton Manufacture of the United States*, 1–16.
41. *Niles' Register* (Jan. 10, 1818).
42. [Carey], "At a very numerous and respectable meeting of Farmers and Manufacturers. . . ."; Carey, *(Private) Sir, Although it is*; Tibbits, *Essay on the Expediency and Practicability of Improving or Creating Home Markets*.
43. "Speculator," *Carlisle Republican* (June 8, 1819).
44. *Memoirs of the Society of Virginia for Promoting Agriculture Containing Communications on Various Subjects in Husbandry and Rural Affairs* (Richmond: Shepherd and Pollard, 1818), iii.
45. Garnett, "Defects in Agriculture," 55–57; Garnett, "Address"; Warden, *Statistical, Political, and Historical Account*, vol. 2.
46. Black, "Essay on the Intrinsic Value of Arable Land."
47. *Memoirs of the Philadelphia Society for Promoting Agriculture*, 1818, 1826.
48. Richard Peters writing for the Philadelphia Society for Promoting Agriculture, unpublished statement "by order of the society," Nov. 28, 1818, Philadelphia Society for Promoting Agriculture (PSPA), Records 1785–1992, box 26, folder 448; Richard Peters to Robert Vaux, Feb. 15, 1818, ibid., box 3, folder 150.
49. Butler, *Farmer's Manual*, 72–73; Adams, *Agricultural Reader*, 32–35.
50. The idea of human progress expressed through the improvement of agricul-

ture certainly was not new in the nineteenth century. Improvement only magnified an impulse that can be linked back to the Enlightenment if not earlier. Joyce E. Chaplin, *Anxious Pursuit*, argues that planters of the lower South had assimilated the key ideas of the Scottish Enlightenment by the 1730s and saw no contradiction between slavery and the construction of a modern South.

51. It is ejected "with some force," according to Van Soest, *Nutritional Ecology of the Ruminant*, 182. The stomach occupies three-quarters of the abdominal cavity in true ruminants, including sheep, cattle, goats, deer, and antelope.
52. Ibid., 182–90; Cole and Garrett, *Animal Agriculture*, 447–59.
53. Sinclair, *Code of Agriculture*, 50, 126.
54. Buel, *Address Delivered before the Berkshire Agricultural Society*, 21.
55. Brady, *Nature and Properties of Soils*, 500–1. Some regarded chicken manure as the very best for its remarkable nitrogen content, but cows produced more manure in less time.
56. Ibid., 505–6.
57. Meinert, interview, Oct. 4, 1999; Nieminen, interview.
58. Ibid.
59. Lorain, *Nature and Reason*, 323.
60. Sinclair, *Code of Agriculture*, 127.
61. *Cultivator* 1 (July 1834), quoted from the 2nd ed. (Albany, 1837), 68–69; Nicholson, *Farmer's Assistant*.
62. John Lorain to James Mease, June 28, 1812, printed in *Memoirs of the PSPA*, vol. 3 (1814), 98–102. "Thus we see that manure is an expensive article, whether it be bought, or made on the farm." Lorain, *Nature and Reason*, 334, 323.
63. Buckner, "Rural Economy—No. III"; Sinclair, *Code of Agriculture*, 36–37.
64. Pyne, *Vestal Fire*, 160.
65. Kerridge, *Agricultural Revolution*, 39–40, 181.
66. Prince, "Changing Rural Landscape, 1750–1850," and Jonathan Brown and H. A. Beecham, "Arable Farming," in Mingay, ed., *Agrarian History of England and Wales*, 44–45, 296–98; Allen and Grada, "On the Road Again with Arthur Young," 114; Chambers and Mingay, *Agricultural Revolution, 1750–1880*, 7–10, 54–55; Duncan, *Centrality of Agriculture*, 63–69.
67. Sinclair, *Code of Agriculture*, 50. Kerridge, *Agricultural Revolution*, 311, writes that "the innovations of the agricultural revolution were designed first and foremost to increase the volume and range of fodder and forage crops turned out by English farmers." Farm stock began to improve as a consequence. Ecologists refer to a P/R ratio to measure the balance between production and decomposition in ecosystems. The two are not al-

ways in balance year after year, but a community with a P/R ratio just above or just below 1 will tend to maintain itself over time. Golley, *Primer for Environmental Literacy*, 115.

68. Eisenberg, *Ecology of Eden*, 4–8; Zohary and Hopf, *Domestication of Plants in the Old World*, 32.
69. Warden, *Statistical, Political, and Historical Account*, vol. 1, 525. Poverty grass is a native species of wild oat (sometimes called poverty oat) with low nutritional value. Palmer, *Fieldbook of Natural History*, 353–54.
70. Olmsted, *Walks and Talks of an American Farmer*, 86–87; Fessenden, *Complete Farmer and Rural Economist*, 20.
71. See, for example, "General View of the Agriculture of the County of Suffolk" (1797), in *Review and Abstract of the County Reports to the Board of Agriculture*.
72. Washington, *Letters*, Dec. 5, 1791.
73. Young, *Six Weeks Tour*, 1–24.
74. Ibid., 22.
75. Young, *Political Arithmetic*, 155; Young, *Six Weeks Tour*, viii–x; Young, *General Report on Enclosures*. Young may have been the first to imagine a countryside of fewer and fewer farmers, each producing to great efficiency, while the rest went to cities to become wage earners and consumers.
76. The owners of great houses owed much to parliamentary acts of enclosure and their ancient capital. Or, as Raymond Williams put it, "a form of legalised seizure enacted by representatives of the beneficiary class . . . and in the acreage it affected—a quarter of all cultivated land—it can be said to be decisive." Williams, *The Country and the City*, 98, 107.
77. Prince, "Changing Rural Landscape, 1750–1850," 22–50. Chambers and Mingay, *Agricultural Revolution, 1750–1880*, 35.
78. Sinclair, *Code of Agriculture*, 5.
79. Solbrig and Solbrig, *So Shall You Reap*, 90–92; Duncan, *Centrality of Agriculture*, 53–62.
80. Buel, *Farmer's Companion*, 66; Grisenthwaite, "New Theory of Agriculture"; John Lorain, "On the Agriculture of England," *Memoirs of the PSPA*, vol. 3 (1814), 93. For an example of the juvenile literature on the subject of agriculture, see *William and Eliza; or, The Visit*.
81. *Memoirs of the Pennsylvania Agricultural Society* (1824), 82.
82. Buckner, "Rural Economy—No. III"; Butler, *Farmer's Manual*, 107. Whitaker, *Claims of Agriculture to Be Regarded as a Distinct Science*, claims that convertible husbandry did not work in the southern states.
83. Allen and Grada, "On the Road Again with Arthur Young," 104. The authors argue that Young exaggerated the power of enclosure to raise efficiency. Chambers and Mingay, *Agricultural Revolution, 1750–1880*, 34–36, are more positive about eighteenth-century innovation and enclo-

sure. They say that total agricultural produce increased by 43 percent. In England and Wales, wheat yields rose by 10 percent, from about twenty to twenty-two bushels per acre. Deane and Cole, *British Economic Growth, 1688–1959*.

84. Lewis C. Beck, "Production of Breadstuffs," *De Bow's Review* 7 (Aug. 1849), 151–63.
85. Adams, *Agricultural Reader*, 32–35. A note on agricultural schools: the Rensselaer Institute of Troy, New York, established in 1824, may have been the first technical school in the United States—dedicated to math, physics, chemistry, natural history, civil engineering, manufactures, and agriculture. The only older institution, The Military Academy at West Point, mainly taught engineering.
86. Black, "Essay on the Intrinsic Value of Arable Land"; *Memoirs of the Society of Virginia for Promoting Agriculture*.
87. See Thomas Cole's five-painting series, *The Course of Empire* (in the collection of the New-York Historical Society).
88. Thoreau, *Walden*, 98–99.
89. Thoreau, "Ktaadn," in *Maine Woods*, 94–95.

Two

1. Benton, *Mr. Benton's Speeches on the Public Lands*, 25–45.
2. Thornton, *Cultivating Gentlemen*, 124–25.
3. Cayton, *Frontier Republic*, xi, "Planning the Republic," passim.
4. Marsh, ed., *Life and Letters of George Perkins Marsh*, 28–29. Judd's description of farmland reform evokes Marsh: "The admonitions of farm reformers, aimed at using the land more profitably, were both modern and traditional, utilitarian and spiritual. Concern for farmscape and landscape, bird life, and farm woodlots all dealt with the matter of resource efficiency, yet they carried a heavy moral content." Judd, *Common Lands, Common People*, 89.
5. Grammer, *Pastoral and Politics in the Old South*, 42–43; Bradford, "A Virginia Cato: John Taylor of Caroline and the Agrarian Republic," introduction to *Arator*.
6. Taylor, *Arator*, 79–80.
7. Ibid., 317.
8. Stilgoe, *Common Landscape of America*, 12–13.
9. The Midland frontier culture is Finnish in origin and American Indian in influence, according to Jordan and Kaups in their remarkable book, *American Backwoods Frontier*, 1–14, 233; Faragher, *Sugar Creek*.
10. Meinig, *Continental America, 1800–1867*, 226–27.
11. Klein, *Pennsylvania Politics, 1817–1832*, 1–7; Ben Marsh and Pierce

Lewis, "Landforms and Human Habitat," in Miller, ed., *Geography of Pennsylvania*, 18–21.

12. Estwick, *Pedestrious Tour of Four Thousand Miles*, 44, 24.
13. The best account of emigration from an eastern state remains Barron's *Those Who Stayed Behind*. Also see Flint, *Letters from America*, 65–68, and Keep, *Address to the Agricultural and the Ladies' Manufacturing Societies*, 12. Census information from 1810, 1820, and 1830 taken from United States Historical Data Browser, http://fisher.lib.virginia.edu/census/. Data for this site were collected, keypunched, and error checked by the Inter-university Consortium for Political and Social Research (ICPSR). The selection of data proceeded according to guidelines developed by the American Historical Association (Sept. 28, 1999).
14. Faragher, *Sugar Creek*, 45.
15. Philadelphia Society for Promoting Agriculture, *Address to the Citizens of Pennsylvania on the Importance of a more Liberal Encouragement of Agriculture*, 5; Black, "Essay on the Intrinsic Value of Arable Land." William Tilghman, chief justice of Pennsylvania and a vice president of the Society, could only hope that someone would "convince the emigrants from the eastern states, and from Europe, that it is [in] their interest to establish themselves here, rather than seek a residence in a distant country, to the west or the south." Tilghman, *Address Delivered before the Philadelphia Society for Promoting Agriculture . . . January 18, 1820.*
16. John Hare Powel, a cattle breeder, became so frustrated by endless addresses and stillborn schemes that he bolted the PSPA to found the Pennsylvania Agricultural Society. He said privately what others had kept to themselves: that the efforts of the PSPA "are not calculated to stimulate the practical husbandmen of this state." The bylaws of Powel's club prevented any meeting within ten miles of city limits; all the officers were to be "practical farmers." Practical farmers stayed away, so the new society simply replicated the manner and membership of the older. John Hare Powel to Charles Miner, Aug. 30, 1822, Philadelphia Society for Promoting Agriculture, Records 1785–1992, box 3, folder 163, Van Pelt Library, University of Pennsylvania.
17. Kulikoff, *Agrarian Origins of American Capitalism*, 218.
18. "Pennsylvania Farmers," *Southern Agriculturist* (Jan. 1828); for more on land prices during the 1820s, see Flint, *Letters from America*, 67; Howitt, *Selections from Letters*.
19. Caleb Kirk, "Account of the uncommonly fertile soil of Lancaster county," *Memoirs of the PSPA* (1818), 82–83.
20. Bowen, *Pictorial Sketch-Book of Pennsylvania*, 31; Fletcher, *Pennsylvania Agriculture and Country Life, 1640–1840*, 123.

21. Fearon, *Sketches of America*, 184. From another English observer impressed by Lancaster: "I had heard much of this county, but the land as far surpassed my anticipation as Susquehanna county fell short of it. After travelling over upwards of 2000 miles in different states, I do not hesitate to say that the land from Ebensburgh hither, and round this place, is the finest I have seen. A farm, including all its varieties of soil, is worth 120 to 150 dollars per acre. . . . The farm houses all along resembling gentlemen's seats in England." Howitt, *Selections from Letters*, 205–7; Flint, *Letters from America*, 67.
22. Cobbett, *Year's Residence in the United States of America*; Herman, "The Model Farmer and the Organization of the Countryside," 54.
23. Pierce Lewis, "The Northeast and the Making of American Geographical Habits," in Conzen, ed., *Making of the American Landscape*, 95; Ensminger, *Pennsylvania Barn*, passim.
24. Morris diary (photocopy). For another diary of a prosperous Pennsylvania farmer, see Price's diary of 1831 to 1847.
25. Morris diary. There is a marvelous description of just such a space in the writing of Susan Fenimore Cooper. She came upon a recent clearing with burned-over stumps near her home in Cooperstown, New York, and predicted that the place would soon be plowed: "It frequently happens that land is cleared of the wood, and then left in a rude state, as wild pasture-ground; an indifferent sort of husbandry this, in which neither the soil nor the wood receives any attention; but there is more land about us in this condition than one would suppose." Cooper, *Rural Hours*, 144–50.
26. Fearon, *Sketches of America*, 184; Flint, *Letters from America*; James T. Lemon's estimates for the size of an average farm in the southeast are more modest. In 125 acres a farmer might work 26 acres in arable, 20 in hay and meadow, 5 in flax/hemp and vegetables, and the rest in woods. Lemon, *Best Poor Man's Country*, table 27.
27. Morris diary, Jan. 1823 to Jan. 1824.
28. *Memoirs of the Pennsylvania Agricultural Society* (1824), 42; Sinclair, *Account of the Systems of Husbandry of Scotland*; Abbot, *Address delivered before the Essex Agricultural Society*, 8.
29. Wiggins, *American Farmer's Instructor*, 268. I am certainly not the first to notice this: "Northern farms had a larger number of animal units to the cultivated acres and a greater quantity of manure than the depleted farms and plantations of the South." Gates, *Farmer's Age*, 347. Northeastern owners of 160-acre farms in 1860 owned an average of 13.9 cows and beef cattle and 16.6 sheep, more in both categories than farmers in the Midwest, who owned 8.1 cows and beef cattle and 7.7 sheep on the same-size farm. Atack and Batemen, *To Their Own Soil*, table 8.7, 141. According to Dun-

can, *Centrality of Agriculture*, 65, the acreage dedicated to wheat in rural England "was probably something like a third or a quarter of the total acreage being farmed," with the rest in grass and fodder crops.

30. The following table describes neat cattle and hay in the Northeast, Northwest, Southeast, and Southwest, with raw numbers, density per square mile, and percentage of total production in the United States. Source: Census of 1840.

Region	Area in square miles	Head of cattle	Animals/ sq. mi.	% of total U.S. cattle	Tons of hay	Tons/ sq. mi.	% of total U.S. tons
Northeast	164,230	4,903,267	29.86	33	7,879,836	47.98	77
Northwest	368,315	3,151,511	8.56	21	1,593,777	4.33	15
Southeast	210,209	3,324,255	15.81	22	614,351	2.92	6
Southwest	328,197	3,589,279	10.94	24	158,862	0.48	2
Total	1,070,951	14,968,312	13.98	100	10,246,826	9.57	100

31. Judd, *Common Lands, Common People*, passim.
32. *Carlisle Republican*, July 6, 1819; Morris diary, May, July, Oct. 1823.
33. Morris, *Address Delivered before the Agricultural Society of Bucks County*, 9; William Ellery Channing, *Self-Culture* (1844), 8–9, 24–25, quoted in Davis, ed., *Antebellum American Culture*, 69.
34. Morris, *Address Delivered before the Agricultural Society of Bucks County*, 12.
35. Ibid., 6.
36. Rush, "An Account of the Progress of Population, Agriculture, Manners, and Government in Pennsylvania, in a letter to a friend in England," in *Essays*, 213–25; Herman, "The Model Farmer and the Organization of the Countryside," 54. Tatham had a different take on the same kind of social progression: after the opening and the improvements the land is worn out, when "it becomes an object to the *German farmer*, who improves it by a system of European husbandry." Tatham, *Communications concerning Agriculture and Commerce*, 49–50.
37. Goodwin, *Address Delivered before the Worcester Agricultural Society*, 5–6.
38. *Proceedings of the Pennsylvania Agricultural Society with the Address of their President Jonathan Roberts, Esq.*, 7–8. Thornton makes a similar point in *Cultivating Gentlemen*, 125. Massachusetts had failed utterly to influence the nation from the election of Thomas Jefferson to the War of 1812. Its agricultural erosion was sure to make its elite farmers even more marginal.
39. Rossiter, *Conservatism in America*, 27–29.
40. Colman, *On Labor*, 23. Or, as Barron puts it, agricultural leaders "advised

contentment with a competence or a modest independence and berated those who went after greater or quicker returns," usually in the West. Barron, *Those Who Stayed Behind*, 36.

41. Buel, *Farmer's Companion*; Jesse Buel to John Armstrong, Dec. 5, 1822, Natural Science Manuscript Group, group 583, box 1; Carman, ed., *Jesse Buel, Agricultural Reformer*, introduction.
42. Buel, *Address Delivered before the Berkshire Agricultural Society, October 5th, 1837*, 22. Buel, *Farmer's Companion*, 21.
43. Howe, *Political Culture of the American Whigs*, 28–42, 101.
44. Lowenthal, *George Perkins Marsh, Versatile Vermonter*, 42. The quotation is from Marsh, *Man and Nature*, 46; Dorman, *Word for Nature*, 14–22.
45. Thomson, *Tradesman's Travels*, 78–80; Lorain, *Nature and Reason*, 315. Within a few decades of settlement new western states replicated the ratio of landed to landless that prevailed in the old states: Henretta, "Families and Farms," 9; Barron, *Those Who Stayed Behind*, chap. 2, passim; Nicholson, *Farmer's Assistant*, 169.
46. Destutt de Tracy, *Treatise on Political Economy*, 53, 124; see McCoy's discussion of the same work in *Elusive Republic*, 250; Kulikoff, *Agrarian Origins of American Capitalism*, 34–47.
47. On the "conservative ethic" of the eastern states, see Welter, *Mind of America, 1820–1860*, 105, 298.
48. *Register of Debates in Congress*, 21st Cong., 1st Sess., Jan. to May 1830, 22–40. For the earlier discussion of the same, see *Mr. Benton's Speeches on the Public Lands*, 39–40. Richard Rush ran for Vice President with John Quincy Adams against Andrew Jackson in 1828.
49. In fact, Webster had long been a friend of western interests. His battle with Hayne and Benton was misinterpreted in New England as a rebuke to western development when Webster never intended that. In the debate itself, Webster never answers Benton's insults and insinuations, but rather attacks Hayne's sectional gambit—a much more offensive stance to Webster. See Parish, "Daniel Webster, New England, and the West."
50. George McDuffie established a model plantation in the Abbeville District in 1821 and became governor of the state during the 1830s. A journalist visited him in 1848 and reported in *De Bow's Review* 6 (Aug. 1848), 149. Also see Feller's excellent study, *Public Lands in Jacksonian Politics*, 84–87, 113–17. *Speech of Daniel Webster in Reply to Mr. Hayne*, 24.
51. Fearon, *Sketches of America*, 193–94.
52. Lorain, *Hints to Emigrants*, 26.
53. Linn, *History of Centre and Clinton Counties, Pennsylvania*, 394; *Commemorative Biographical Record of Central Pennsylvania*, 176, quoted from a typescript photocopy in the Pennsylvania State University Library; Lorain, *Hints to Emigrants*, 18; Lorain, *Nature and Reason*, 408.

54. *Illustrated Souvenir History of Philipsburg, Pennsylvania*; Melish, *Map of Pennsylvania*; Federal Writers' Program of the Works Projects Administration, *Pennsylvania*, 467; Marsh and Lewis, "Landforms and Human Habitat," fig. 2.8, in Miller, ed., *Geography of Pennsylvania*.
55. Warden, *Statistical, Political, and Historical Account*, vol. 1, 103.
56. Pearson, *Notes during a journey*, 48.
57. Ibid., 59.
58. Birkbeck, *Letters from Illinois*, 15–19, 90–91.
59. Lorain, *Hints to Emigrants*, 15–16.
60. Cobbett, "Letter to Morris Birkbeck, Esq.," in *A Year's Residence in the United States of America*, 301–2, 308–10, 395.
61. Lorain, *Nature and Reason*, 168, 185–86.
62. Ibid., 130–31, 336–41.
63. Jordan and Kaups, *American Backwoods Frontier*, 1–14, 233, 253; Lorain, *Nature and Reason*, 333.
64. Lorain, *Nature and Reason*, 488, 111.
65. Ibid., 403–10, 423–24.
66. Ibid., 410.
67. Ibid., 519, 336, 503.
68. Livingston, *Essay on Sheep*, 20–21.
69. Ibid., 6–8.
70. Board of Agriculture, State of New York, *Memoirs of the Board of Agriculture*, vol. 1 (1821), 307; United States Bureau of Animal Industry, *Special Report on the Sheep Industry* (1892), quoted in Bidwell and Falconer, *History of Agriculture in the Northern United States, 1620–1860*, 218. Through still more diplomacy, E. I. du Pont acquired the full-blooded ram Don Pedro in a deal that also involved Thomas Jefferson. The profligate ram created the original stock behind Du Pont & Company of Wilmington, Delaware. For an early reference to the Du Pont company, see *Cultivator* (Feb. 1835).
71. *Merino Sheep*.
72. *Niles' Weekly Register* (Dec. 31, 1814); Humphreys, *Discourse on the Agriculture of the State of Connecticut*; Buckner, "Rural Economy—No. III."
73. "Sheep Husbandry," *Cultivator* 1 (Jan. 1835); "Dry pasture, fallow fields, and lands covered with stubble afford the best nourishment for sheep." *American Republican* (March 25, 1828); "Hints on Sheep Husbandry," *Cultivator* 1 (Jan. 1835); *Constitution of the Merino Society*, 5. Another investor proposed that five thousand sheep could be kept on one thousand acres. Dr. Robert H. Rose, "Estimate of the probable Profit in keeping one thousand Sheep," *Memoirs of the PSPA*, vol. 3 (Philadelphia: Johnson and Warner, 1814), 355. A Pennsylvania farmer said that sheep "will turn

thorns, and briars, and noxious weeds . . . into useful substances, and in the end cause them to become extinct, and valuable grasses to grow in their stead." *Annual Report of the Commissioner of Patents for the year 1848*, 13th Cong., 2nd sess., H. Exec. Doc. 59, 459. Bard, *Guide for Young Shepherds*, v. Said the members of a typical convention: "The raising of sheep gives value to lands not suited to ordinary cultivation, and makes worn-out fields productive of profit to the farmers, if wool fetches a reasonable price." *General Convention of Agriculturists and Manufacturers* [also known as "the Harrisburg Convention"], 64–65.

74. Ryder, *Sheep and Men*, 425–36; Board of Agriculture, State of New York, *Memoirs of the Board of Agriculture*, vol. 1 (1821), 307.
75. Bard, *Guide for Young Shepherds*, 31–38; *Cultivator* 1 (Aug. 1834).
76. *Journal of a Convention of Delegates of the State of Pennsylvania*, 7–9.
77. Benton and Barry, *Statistical View of the Number of Sheep*.
78. Stilwell, "Migration from Vermont," 172–74; *Journal of a Convention of Delegates of the State of Pennsylvania*, 7–9; Barron, *Those Who Stayed Behind*, 21–22. In Vermont, sheep declined in importance, along with wool prices, after the Civil War.
79. Bidwell and Falconer, *History of Agriculture in the Northern United States, 1620–1860*, 219; "Domestic Industry," *American Farmer* 1 (Sept. 24, 1819).
80. Goodrich, *Recollections of a Lifetime*, 404–5; Board of Agriculture, State of New York, *Memoirs of the Board of Agriculture*, vol. 1, 308. The Agricultural Society of Pendleton, South Carolina, called people who bought sheep on speculation "merino mad"—"many have learned by experience, that this was a delusion, and now know, that a flock of sheep larger than is necessary for domestic consumption, is a useless expense." *American Farmer* (Oct. 5, 1821); and see Lowell, "Address Delivered before the Massachusetts Agricultural Society."
81. Richard Peters, "Notices for a Young Farmer; Particularly ones on Worn Lands," *Memoirs of the PSPA*, vol. 4; Lowell, "Address Delivered before the Massachusetts Agricultural Society," 230–31. Said Lowell: "A man must be wholly ignorant of the history of Great Britain, who does not know, that to the products and the improvement of her sheep, she has been indebted for the largest portion of her agricultural and commercial power."
82. *Constitution of the Merino Society*, 4–5.
83. Census of 1820. A census of 1837 found 24 establishments in Maine, 43 in New Hampshire, and 70 in Vermont. But Massachusetts and New York dominated, with 519 and 234 companies, respectively.
84. These sentences are based on the work of Brooke Hunter from a paper titled "Grist for the Mill: The Refinement of an American Industry," part of

her forthcoming Ph.D. dissertation at the University of Delaware. For the New England model of industrial development, see Prude, *Coming of Industrial Order*.

85. They would have agreed with Jesse Buel: "That an improved state of agriculture in any country, is most favorable to manufactures and commerce is such a truism, as needs no setting forth or illustration." Jesse Buel to John Armstrong, Dec. 5, 1822, Natural Science Manuscript Group, group 583, box 1; *Memorial of the Ontario Agricultural Society*; Carey, *New Olive Branch*.
86. *Proceedings of the Pennsylvania Agricultural Society with the Address of their President Jonathan Roberts, Esq.*, 12–13; Tibbits, *Essay on the Expediency and Practicability of Improving or Creating Home Markets*. For news of the vote in the state legislature, see *Berks and Schuylkill Journal* (Feb. 3, 1827).
87. *General Convention of Agriculturists and Manufacturers*.
88. [Carey?], *To the Honourable*, 2–3. It is unclear how many of these statements came from Carey's pen. He argued unceasingly that only manufacturing could provide a stable market for agriculture. *Memorial of the Ontario Agricultural Society; Digest of Accounts of Manufacturing Establishments*.
89. *American Republican* (Feb. 5, 1828); "Tariff Bill," in *Register of Debates in Congress*, 20th Cong., 1st sess., March 5, 1828, 1757. One representative from Pennsylvania considered the interests of wool growers and manufacturers in the same breath, as a unity of interests, calling both "men who have been engaged for many years in these, at present, unprofitable branches of industry, and are now humbly supplicating this body for such protection as will save them from bankruptcy and ruin." Speech of Mr. Anderson of Pennsylvania on the Tariff, Delivered in the House of Representatives, March 27, 1828, in *Village Record* (Chester, Pa.), April 23, 1828.
90. [Carey?], *To the Honourable*, 5; *Journal of a Convention of Delegates of the State of Pennsylvania*, 7–9.
91. Faux, *Memorable Days in America*, 133; Wright, *Wool-Growing and the Tariff*, 50–53; Tibbits, *Essay on the Expediency and Practicability of Improving or Creating Home Markets*, 3–9; *Memorial of the Agricultural Society of South Carolina; Resolutions and Remonstrance of the Agricultural Society of St. John's, Colleton, S.C.*
92. There is much more to it than that. According to Freehling, *Prelude to Civil War*, 29–36, planters had not only suffered the economic travails of the 1820s; they had seen the rise of antislavery. A great shift had taken place in their national stature between 1816 and 1832. Freehling acknowledges that a decline in the productive capacity of land, especially in the upcountry, played a part in the coming of the crisis. Cauthen, "Secession Movement before 1860," 133.

93. For a vigorous statistical treatment of the decline of the low country in particular, see Coclanis, *Shadow of a Dream*. As Coclanis states, "The economy of the South Carolina low country collapsed in the nineteenth century. Collapse did not come suddenly—many feel, for example, that the area's 'golden age' lasted until about 1820—but come it did nonetheless." The author makes almost no mention of ecological factors tied to agriculture as a cause of the collapse.
94. Ruffin, *Agriculture, Geology, and Society in Antebellum South Carolina*, 76–78.
95. Ibid., 100, 126.
96. For much more on labor and the formation of a southern landscape, see Stewart, *"What Nature Suffers to Groe."*
97. From the finding aid to the Law Papers. Thomas Cassells Law (1811–1888), Farm Journal, 1841–1842, in the Law Papers, 1810–1865, Darlington District, South Carolina; Records of Ante-bellum Southern Plantations from the Revolution through the Civil War, series A, part 2.
98. Mills, *Statistics of South Carolina*, 517–19.
99. Law, Farm Journal, March 1844 and March 1845. Robert Mills noticed that "immense tracts of excellent land along the margins of the creeks lie uncultivated and unreclaimed, though they provide fine ranges for cattle all the year round." Mills, *Statistics of South Carolina*, 518; "Report on Cotton, to the Newbury [*sic*] Agricultural Society (1847?)," in Turner, ed., *Cotton Planter's Manual*, 47.
100. "Culture of Cotton," *Southern Agriculturist*, n.s. (Sept. 1841), 462.
101. "Report of the Committee Appointed by the Agricultural Society of Cambridge, (S. C.) on Cotton," *Southern Agriculturist*, n.s., 1 (Dec. 1841); "Extracts from an Address, delivered before the Greenville Agricultural Society, August 1841," in *Southern Agriculturalist*, n.s., 2 (Jan. 1842). Said Edmund Ruffin on the plan of penning cattle during the summer on nothing but forest litter, "However it may add to the quantity and quality of the manure, [it] is very injurious to the cattle." Ruffin, *Report of the Commencement and Progress of the Agricultural Survey of South Carolina*, 83.
102. "On the Causes of Emigration," *Southern Agriculturist* (Nov. 1839). The practice of putting down manure for small grains but not for cotton was very common. See Hall, *Story of Soil Conservation*, 15. Hall writes, "It was feared that rotations and green manuring would divert too much labor from the all-important cotton and corn" (19). N. B. Cloud is a good illustration of a planter who assembled almost all the elements of the northern system, and he related the common reaction of planters when they saw or heard of his methods: "Sir, your systems are beautiful, your exertions are praiseworthy, all your manure-making . . . are conditions actually essential to the improvement of our agriculture; but . . ." "Dr. N. B. Cloud's Improved Sys-

tem of Cotton Culture," in Turner, ed., *Cotton Planter's Manual*, 70–78. Mississippi quotation from Moore, *Emergence of the Cotton Kingdom*, 36.

103. Rubin, "Limits of Agricultural Progress."
104. Tatham, *Communications concerning the Agriculture and Commerce of the United States of America*, 66; "On the Causes of Emigration," *Southern Agriculturist*, 357.
105. Mills, *Statistics of South Carolina*. Edmund Ruffin, flustered by the backwardness he saw on a visit to Edgecombe County, North Carolina, during the 1850s, described the scarcity of hay "owing to the wide extent of cotton culture, and the small extent of forage crops and products—and the entire want of grass culture and of meadows. . . . To supply the deficiency, *northern* hay is imported." Ruffin, *Agricultural, Geological, and Descriptive Sketches of Lower North Carolina*, 295–96; Ruffin, *Agriculture, Geology, and Society in Antebellum South Carolina*, 21. Fence laws also had something to do with the general weakness of convertible husbandry in the region. In places with abundant woodland, like the Piedmont until about the middle of the century, laws encouraged planters to own more stock than they might have had they been required to fence them. See Hahn, *Roots of Southern Populism*, 60–63.
106. Meinert, interview, Dec. 7, 1999; Tatham, *Communications concerning the Agriculture and Commerce of the United States of America*, 66.
107. Law, Farm Journal, Feb. and April 1843, Law Papers.
108. Kirby, *Poquosin*, 118–19, offers a similar defense: "Was there not ultimate wisdom, perhaps, in rustic carelessness and impersistence, and in a willingness to live close to animals in shelters less than grand?"
109. Stirling, *Letters from the Slave States*, 178.
110. For these matters as they relate to Mississippi, see Moore, *Emergence of the Cotton Kingdom*, 33–36.
111. This paragraph owes its inception to Earle's "Myth of the Southern Soil Miner."
112. "Col. Chappell's Address," 12–17.
113. J. Hamilton Couper, "Essay on Rotation of Crops," *Southern Agriculturist* (Feb. 1833), 65–66; Stewart, *"What Nature Suffers to Groe,"* 123, 186–88; "Col. Chappell's Address," 12–17.
114. There can be no doubt of their goals. They dreamed of forty or fifty bushels of grain per acre, pine barrens converted into "fruited fields," and cattle at a thousand pounds "or more." *Papers Published by Order of the Agricultural Society of South Carolina*, 6–8.
115. "Observations on the Mode of Renovating Soils," *Southern Agriculturist*, n.s. (May 1841), 233.
116. Law, Farm Journal, June 1843, Law Papers. The affliction, probably nut

grass, could have come from a lack of fertility or a high acid content in the soil. Seeds in the soil from long before clearing germinated under conditions of high acidity. Law probably failed to apply marl (lime) or to manure his land, which weakened the plants and created the perfect environment for weeds. Meinert, interview, Dec. 7, 1999. Or as one contemporary put it: "In the cotton crop [low fertility] is manifested by a general want of vigour in the plant, a tendency to rust, and a great liability to attacks from lice &c," in "Observations on the Mode of Renovating Soils," *Southern Agriculturist*, 231. The author, "C," was probably John C. Calhoun.

117. N. B. Cloud gave each of his slaves twenty acres, but that was divided up into cotton (five acres), corn (ten acres), and fallow (five acres); Cloud in Turner, ed., *Cotton Planter's Manual*, 77. A writer for the Agricultural Society of South Carolina recommended three and a half acres per slave, calling it "sufficient for a negro to attend, if he, at the same time, raises provisions for his sustenance." Agricultural Society of South Carolina, *Original Communications Made to the Agricultural Society of South Carolina*, 42.
118. "On the Causes of Emigration," *Southern Agriculturist*, 572. "As the land exhausts, an effort is made to equalize the deficiency of yield, by cultivating an increased quantity . . . until at last disheartened, the proprietor has to choose between concentration and emigration. The latter is more frequently resorted to"; quoted in "Observations on the Mode of Renovating Soils," *Southern Agriculturist*. *Soil of the South* 1 (Jan. 1851) called exhaustion a threat to slavery: "He who lives to see the time, when the fertility of our soil shall have been exhausted, will also see the other great basis of Southern wealth crumble."
119. "On the Causes of Emigration," *Southern Agriculturist*.
120. Thomas C. Law, *Report on the Best System of Farming for Improving Our Land*, typescript address delivered for the Hartsville Farmers' Club, Hartsville, S.C., June 8, 1871, Law Papers.
121. "On Manures," *Southern Agriculturist* (March 1833), 121.
122. Coclanis, *Shadow of a Dream*, 134–40.
123. Law, Farm Journal, 1841–1842, Law Papers.
124. Drayton, *View of South-Carolina*, 10–11. Drayton wrote that the Sand Hills consisted of hilly land where detritus from higher locations often collected to form deep soils. He also called it "capable of producing from fifty to seventy bushels of Indian corn, and twelve hundred weight or more of cotton in the seed, to each acre."
125. United States Census, quoted in Hall, *Story of Soil Conservation*, 4; Bennett, *Soil Erosion*, 6.
126. To be exact, these soils exist in the B horizon in random places throughout

the Piedmont, like Greene County in Georgia and Fairfield, Union, and Chester Counties in South Carolina. Trimble, "Man-Induced Soil Erosion," 16n.

127. Drayton, *View of South-Carolina*, 10–11; Michaux, *Travels to the Westward of the Alleghany Mountains*; Klein, *Unification of South Carolina*, 9–15, 246–53.
128. Mills, *Mills' Atlas of the State of South Carolina.*
129. Both quotations from sources quoted by Trimble in "Man-Induced Soil Erosion," 52–55.
130. "Letter from J. R. Poinsett on the Saluda River near Greensville, SC," in *Transactions of the New York State Agricultural Society*, vol. 4 (Albany: E. Mack, 1845), 340; Edmund Ruffin called the region "fine land when new, but soon washed & worn out." Ruffin, *Agriculture, Geology, and Society in Antebellum South Carolina*, 284.
131. United States Soil Conservation Service, USDA, *Soil Survey—Newberry County, South Carolina*, 1. The Piedmont soils of Pennsylvania are also susceptible to erosion, and soil maps of Chester and Delaware Counties from the 1960s show evidence of it. But no source from the first half of the nineteenth century mentions erosion on these lands. United States Soil Conservation Service, USDA, *Soil Survey—Chester and Delaware Counties, Pennsylvania.*
132. United States Soil Conservation Service, USDA, *Soil Survey—Newberry County, South Carolina*, series 1956, no. 10; series 1958, no. 25; series 1958, no. 21; series 1962, no. 10. A study led by Howard Odum estimated that South Carolina had lost 3 million acres and that "of Alabama's approximately 9,100,000 acres in cultivation in 1930, approximately 4,000,000 have been largely denuded of the top soil." Odum, *Southern Regions of the United States*, 339.
133. Brady, *Nature and Properties of Soils*, 438.
134. Ibid., 438; *National Atlas of the United States of America.* One soil at State College, Pennsylvania, had an inherent erodibility (that's the *K* factor) of 0.31—higher than the same measure at Clemson, South Carolina, in the heart of the Piedmont (0.28). The *erodibility* of soils is explained not only by soil type but also by rain erosivity and topography. Hudson, *Soil Conservation*, 27, 42, 62 graph.
135. *Annual Report of the Commissioner of Patents for the year 1848*, 436–54, 477; *Annual Report of the Commissioner of Patents for the year 1847*, 13th Cong., 1st sess., H. Exec. Doc. 54, 385.
136. Harris, *Piedmont Farmer*, Feb. 16, 1859.
137. Ibid., April 8, 1857. "Again the cowfood is nearly out and nothing in the woods for the cattle to eat." March 27, 1858; Slaves "hauling leaves to the horse lot." March 19, 1859; Jan. 28, Oct. 26, 1858.

138. Brady, *Nature and Properties of Soils*, 505.
139. For four years running a sentence like this one appeared in Harris's diary: "We are nearly done clearing newground." Jan. 18, 1859; "Observations on the Mode of Renovating Soils," *Southern Agriculturist*, 225.
140. Harris, *Piedmont Farmer*, Sept. 7–Nov. 29, 1856; May 10, 1857.
141. *Papers Published by Order of the Agricultural Society of South Carolina*, 9–10; Harris, *Piedmont Farmer*, Feb. 20, 1858.
142. I run the risk of making planters seem more rational than they were, since they held an aristocratic conception of themselves in which slaves, as well as poor whites, were essential to their political identities. Genovese, "The Limits of Agricultural Reform," in *Political Economy of Slavery*, passim.
143. "Cheap Farming," *Cultivator* 1 (March 1834); *Farmers' Register* 1 (Aug. 1833), 187. Some did speak against emigration. Said A Pine Land Planter, "I cannot but regard the whole scheme of Western fortune hunting as a splendid delusion." *Southern Agriculturist* (July 1839), 356.
144. Mills, *Statistics of South Carolina*, 516; Thomas Cassells Law, Journal of Travel to the Western Countries from September 2 to October 9, 1834, Sept. 13, 1834, typescript in Law Papers.
145. Thomas Cassells Law, "Report on the Best System of Farming for Improving Our Lands," delivered to the Hartsville Farmers' Club, June 8, 1871, typescript in Law Papers.
146. Edward Means to Mary Hart Means, Nov. 24, 1846, Means Papers. Letter of Nov. 16, 1836, Gilliland Papers, Wilcox County, Alabama, and Charleston, South Carolina. Both collections in the Records of Ante-bellum Southern Plantations from the Revolution to through the Civil War, series A, part 2.
147. Hammond, *Secret and Sacred*, 81; Faust, *James Henry Hammond and the Old South*, 109, 128.
148. Hammond, "Anniversary Oration," 182–83.
149. *Niles' Weekly Register* (Nov. 21, Oct. 24, Aug. 29, 1829; Jan. 19, June 1, 1833; April 19, 1834). "There is considerable emigration from North Carolina, and a great one from South Carolina, to Alabama and Mississippi." *Farmers' Register* 1 (Aug. 1833), 187. *New-York Tribune* (July 21, 1854).
150. *Seventh Census of the United States: 1850*.
151. *Annual Report of the Commissioner of Patents for the year 1848*, 92; Clark and Guice, *Frontiers in Conflict*, 167. The total number of seats in 1870 was 243; see Martis, *Historical Atlas of United States Congressional Districts*.
152. "Embanking and Draining Our Low Lands, to Prevent Emigration," *Southern Agriculturist* 11 (1838), 13–21, quoted in Smith, *Economic Readjustment of an Old Cotton State*, 37–38.
153. "Communication of Dr. Jos. Johnson, to the Agricultural Society of South-

Carolina, on the Improvement of Soils by Marl and Lime," *Farmers' Register* 6, no. 11 (1838), 689.

154. "Extracts from an Address, delivered before the Greenville Agricultural Society, August 1841"; also see *Southern Agriculturist* 12 (Nov. 1839), 70, and 12 (July 1839) for a planter speaking out against emigration.
155. Simms, *Richard Hurdis*, 15, 65–67.
156. Simms, *Social Principle*, 42–43.
157. Ibid., 43.
158. Carey, *Harmony of Interests*, 166–67; *The Plough, the Loom, and the Anvil* 1 (1848); Coclanis, *Shadow of a Dream*, 148.
159. Mathew, *Edmund Ruffin and the Crisis of Slavery in the Old South*. Kirby's *Poquosin* places Ruffin's life in the context of rural landscape. Also see Kirby's edited volume of Ruffin's writings, *Nature's Management*. Other works include Craven, *Edmund Ruffin, Southerner*, and Allmendinger, *Ruffin*.
160. Mathew, *Edmund Ruffin and the Crisis of Slavery in the Old South*, 78–83; Brady, *Nature and Properties of Soils*, 228–29.
161. "Means of Inducing Fertility," *Cultivator* (Nov. 1834); Kirby, introduction to *Nature's Management*.
162. Liebig, *Die organische Chemie in ihrer Anwendung auf Agricultur und Physiologie*, 60–62.
163. Liebig, *Relations of Chemistry to Agriculture*, 59–62; for a useful explanation of agricultural chemistry and Liebig's theories, see Rossiter, *Emergence of Agricultural Science*, 10–28.
164. Ruffin, *Essay on Calcareous Manures*, 16, 13–15. See the first proposition at the beginning of chapter 3.
165. Ruffin, *Report of the Commencement and Progress of the Agricultural Survey of South Carolina*, index.
166. Ibid., 72–73.
167. Edmund Ruffin, "An Address on the Opposite Results of Exhausting and Fertilizing Systems of Agriculture" (1852), in Kirby, ed., *Nature's Management*, 323–44.
168. Ruffin, *Agricultural, Geological, and Descriptive Sketches of Lower North Carolina*, 288–93.
169. Hammond, *Address Delivered before the South-Carolina Institute*, quoted in Ruffin, *Agriculture, Geology, and Society in Antebellum South Carolina*, 15.
170. *Proceedings of the South Carolina Agriculture Society*, published in *Southern Agriculturist* (Aug. 1842), 394–97.
171. Ruffin, *Agriculture, Geology, and Society in Antebellum South Carolina*, 61, 78.
172. *Southern Cultivator* 11 (July 1853), from Mathew, *Edmund Ruffin and the Crisis of Slavery in the Old South*, 108, 89.

173. "Agriculture of South Carolina," *Niles' Register* 64 (March 18, 1843). The region from the tidewater to the Fall Line did indeed have an abundance of marl, extending in a band one hundred miles wide all the way to the gulf.
174. Ruffin made this kind of statement in many places. He recommended one hundred bushels per acre on ten acres as an experiment but two to three hundred bushels per acre typically, depending on the quality of marl. *Farmers' Register* 6 (December 1838).
175. Ibid.; *Southern Agriculturist* (June 1841).
176. *Southern Agriculturist* (June 1841).
177. Dated Silver Bluff, Nov. 13, 1841; Hammond, *Secret and Sacred*, 81.
178. "Marling in South Carolina," *Farmers' Register* 10 (July 1842); James Henry Hammond to Edmund Ruffin, July 7, 1844, quoted in Faust, *James Henry Hammond and the Old South*, 128.
179. Brady, *Nature and Properties of Soils*, 235.
180. Wiggins, *American Farmer's Instructor*, 286.
181. "Use of Lime as Manure," *Southern Agriculturist*, n.s. (June 1841); Freehling, *Prelude to Civil War*, 34.
182. Mathew, *Edmund Ruffin and the Crisis of Slavery in the Old South*, 206–8.
183. Hammond had appointed Ruffin without first asking him, so Ruffin insisted on a limit to his term and resigned with the survey incomplete. The legislature had already appropriated the full amount necessary for a two-year survey. As Ruffin explained in a letter to the *American Agriculturist* 4 (May 1845), "My friend, M. Tuomey, ESQ., was appointed to fill the vacated office, and he completed the unexpired year of my appointed term." "Report on the Geology of South Carolina," a review in *Southern Quarterly Review* 16 (1850), 161; Tuomey, *Report on the Geology of South Carolina*.
184. Hugh Hammond Bennett, a soil scientist with the Bureau of Chemistry during the 1920s, reported the following in *Soil Erosion a National Menace*, 8, a report on the southern states dated 1928: "In the great cotton-producing section of central Texas, known as the black waxy belt, white spots representing exposures of the basal chalk and marl beds . . . dot the landscape. . . . These exposures represent the products of erosion—nonarable land that has been substituted for some of the most productive cotton soil of the world."
185. Hammond, "Southern Industry."
186. Ibid., 26–32.
187. Rhett, *Who Is the Producer?* 1–7.
188. Hammond, *Secret and Sacred*, Sept. 22, 1848, 190–91; "Report on the Geology of South Carolina," *Southern Quarterly Review*.
189. "Is Southern Civilization Worth Preserving?" *Southern Quarterly Review* 3 (Jan. 1851); William J. Cooper, Jr., "Advent of the Territorial Question," in *The South and the Politics of Slavery, 1828–1856*; Hietala, *Manifest Design*, 252.

190. Roberts, *American Alchemy*.
191. Tuomey, *Report on the Geology of South Carolina*, 104–32, 97, 282–83. In the preface to a draft report dated 1844, Tuomey wrote to Hammond: "I am aware that gold mining operations are considered inimical to those pursuits that are most conducive to prosperity and happiness, and in the geological reports of those States where gold is not found, its absence is often made [a] matter of congratulations."
192. "Report on the Geology of South Carolina," *Southern Quarterly Review*.
193. Ruffin, "Address on the Opposite Results of Exhausting and Fertilizing Systems of Agriculture," 343–44.
194. *Transactions of the New York State Agricultural Society for the Year 1842*, vol. 2 (1843), 218. Cooper, *Pioneers*, 15–16. And see Taylor, *William Cooper's Town*.
195. The fair originated with the cattle show first mounted by Elkanah Watson at Pittsfield, Massachusetts, in 1807. Bidwell and Falconer, *History of Agriculture in the Northern United States, 1620–1860*, 187–88.
196. Board of Agriculture, State of New York, *Memoirs of the Board of Agriculture*, vol. 1 (1821), 52.

THREE

1. *Congressional Globe*, House of Representatives, 30th Cong., 1st sess., 337–40. Also see Lowenthal, *George Perkins Marsh, Versatile Vermonter*, 74, 107–8.
2. Marsh, *Address*, 8.
3. Judd, *Common Lands, Common People*, passim.
4. Marsh, *Address*, 18–19.
5. Buel, *Farmer's Companion*, 55; Lee, "Philosophy of Tillage," 357.
6. *Farmers' Register* 1 (Feb. 1834).
7. Marsh, *Address*, 18–19. Marsh's first wife, Harriet Buell, daughter of Colonel Ozias Buell, may have been a relative of Jesse Buel's.
8. Marsh, *Man and Nature*, 464, 280; George Perkins Marsh to Spencer Baird, May 21, 1860, quoted in Lowenthal, *George Perkins Marsh, Versatile Vermonter*, 248.
9. "Correspondence, Forests and Civilization," a letter from J. B. Harrison, corresponding secretary of the American Forestry Congress, relating a letter to him from Charles Eliot Norton of Harvard University, in *Garden and Forest* (July 10, 1889).
10. Shaler, *Man and the Earth*, 126–29; Pinchot, *Breaking New Ground*, xix.
11. Pinchot, *Breaking New Ground*, 22.
12. Grove, *Green Imperialism*, 5, 150–54, 470–71n. The French Physiocrats

influenced both the tropical-island botanists and English agricultural thinkers like Arthur Young.

13. Pinchot, *Fight for Conservation*, 20.
14. Bailey, *Harvest of the Year*, 206.
15. Nearing and Nearing, *Living the Good Life*; Rodale, ed., *How to Grow Vegetables and Fruits*, 5; Berry, *Unsettling of America*, 47; Gregory Barton, "Sir Albert Howard and the Forestry Roots of the Organic Farming Movement."
16. Jesse Buel, "On the Necessity and Means of Improving Our Husbandry," *Cultivator* 5 (April 1838). Emphasis in original.
17. Schweinitz, *Journey . . . to Goshen*, 233–34; Robinson, *Pioneer and Agriculturist*, vol. 1, 8.
18. "Description of Northwestern Indiana," *Madison Republican and Banner* (Jan. 15, 1835), in Robinson, *Pioneer and Agriculturist*, 46–56.
19. "Advice to Western Emigrants," *American Agriculturist* (Nov. 1840); "To Western Emigrants," *Cultivator* 7 (Oct. 1840); "Proposition to Facilitate Agricultural Improvement," *Cultivator* 5 (May 1838)—all in Robinson, *Pioneer and Agriculturist*, 518, 141–44, 87–88.
20. Wines, *Fertilizer in America*, 43.
21. American Guano Company, *Report of Experiments with American Guano.* "Chincha Guano Islands," *American Agriculturist* 11 (Oct. 26, 1853), 99, describes an island covered with guano "to the depth of one hundred and twenty-five feet. . . . It is still a question with many what this guano is. I cannot believe it is wholly the excrements of the birds." Wines, *Fertilizer in America*, 61; Jay, *Statistical View of American Agriculture*, 56. England consumed 210,000 tons the same year, an increase of 20 percent over the previous year.
22. "Flight through Connecticut Continued," *American Agriculturist* (Nov. 1849), in Robinson, *Pioneer and Agriculturist*, 307–9, 486; Robinson, *Guano*, 9.
23. Greeley, *What I Know about Farming*, 116–19.
24. Guano arrived during a time of growing interest in soil analysis. Rossiter, "Norton and the Craze over Soil Analysis," in *Emergence of Agricultural Science*; F. W. Johnston, "Of the Economical Uses of Bone as a Manure," *Transactions of the New York State Agricultural Society, 1847* (Albany, 1848), 369. The same point is made by Wines in *Fertilizer in America*, 33, 53.
25. Robinson, *Guano*, 5.
26. McClelland, *Sowing Modernity*, 152–54.
27. Ellsworth, *Improvements in Agriculture and the Arts of the United States*, 5–6; Fessenden, *Complete Farmer and Rural Economist*, 330; *Cultivator* 1 (May 1834).
28. Lee, "Philosophy of Tillage," 342–43, 357; Boyd, *History*, 167–68. Also see

Meeker, *Life in the West*, and Jay, *Statistical View of American Agriculture*, 55.

29. Howe, *Adventures and Achievements of Americans*, 157.
30. For more on how intensive cultivation brought about the first industrial countryside in the United States, see Stoll, *Fruits of Natural Advantage*.
31. Robinson, *Guano*, 7–12. Planters were slow to adopt guano, as they had been slow to adopt other soil-conserving measures, but after the Civil War they became the most voracious consumers of commercial fertilizers in the United States. Wines, *Fertilizer in America*, 41–42.
32. For the culture of the New York suburb, circa 1840, see Stilgoe, *Borderland*.
33. "General View of American Agriculture," *Annual Report of the Commissioner of Patents for the year 1849, part II*, 31st Cong., 1st sess., 1849, S. Exec. Doc. 15, 25; Stilgoe, *Borderland*, 69–77.
34. Robinson, ed., *Facts for Farmers*. Robinson begins the book, "Above all are animals valuable to the farmer, because they convert the coarse products of the farm into manure, without which the owner can not produce food for his own sustenance" (13, 515).
35. "Flight through Connecticut" and "Flight through Connecticut, Continued," *American Agriculturist* (Oct. and Nov. 1849), in Robinson, *Pioneer and Agriculturist*, 247, 309–13.
36. For a full discussion of this subject with attention to horticulture and stockbreeding, see Thornton, *Cultivating Gentlemen*, 170–80. These desires began to come out in the late 1830s: "The farm offers a field for the embellishments of taste in the construction of buildings, in the laying out of grounds, in the leading of water courses, in the arrangement of the garden, in the planting of trees, in the cultivation of flowers, so as to combine and embody the highest efforts of the graphic art." Colman, *An Address Delivered October 2, 1839*, 18.
37. Allen, *Rural Architecture*, 348.
38. *Transactions of the New York State Agricultural Society for the Year 1841*, vol. 1 (1842), 86.
39. Downing, *Rural Essays*, 391.
40. Ibid., 391–95.
41. Ibid., 388; Downing, *A treatise on the theory and practice of landscape gardening*, 121; "Westchester County," *Transactions of the New York State Agricultural Society . . . 1847* (1848), 773–75.
42. *Soil of the South* 1 (Jan. and March 1851).
43. Hilgard, *Report of the Geological and Agricultural Survey of the State of Mississippi*, 4–9.
44. Hilgard, *Address on Progressive Agriculture and Industrial Education*, 5–9. Hilgard stated, "Our agriculture is not a correct, a natural, and scientific

one. We exhaust our lands, abandon them, and resort to others; we remedy, by the abundance of our lands, what we ought to prevent by our skill and science, and, after a longer continuation of such an agriculture, our States must be exhausted, and their inhabitants reduced to poverty." Also see Harper, *Preliminary Report on the Geology and Agriculture of the State of Mississippi*, 19.

45. He bought the farm in 1853. Boyd, *History*, 32; Greeley, *What I Know about Farming*, 19–22.
46. Greeley, *What the Sister Arts Teach as to Farming*, 6–7.
47. Greeley, *What the Sister Arts Teach as to Farming*, 9, 30–32; *Mr. Greeley's Letters from Texas and the lower Mississippi*, 11–14; Greeley, *Overland Journey from New York to San Francisco*, 331–33. Greeley said, "I believe firmly in irrigation."
48. Boyd, *History*, 10–16.
49. Rossiter W. Raymond, "Gray's Peak—to It and up It," *Overland Monthly and Out West Magazine* 5 (Dec. 1870); Boyd, *History*, 143–44. The idea of merging social and agricultural goals was not limited to the West. The New England Emigrant Aid Company, inspired by Northern antislavery politics during the 1850s, established a town in Virginia in an attempt to plant a beachhead for the family farm in the plantation South. The critique of Southern agriculture toward the Civil War is pervasive in *The Plough, the Loom, and the Anvil*, edited by John Skinner. See Phillips, "Antebellum Agricultural Reform," 808–9.
50. Boyd, *History*, 268, 127; Mead, "Government Aid and Direction in Land Settlement," 83, 94–96; Mead, *Irrigation Institutions*; Conkin, "The Vision of Elwood Mead," 88–97.
51. Cooper, *Forest Culture and Eucalyptus Trees*, 9–11; Tyrrell, *True Gardens of the Gods*, 62–65.
52. See Kirby, *Poquosin*, 118, where Kirby asks the same question.
53. Mumford, *Pentagon of Power*, 134–53. Mumford had written earlier on the subject in *Technics and Civilization* (1934) but had been inspired to revisit it by "the wholesale miscarriages of megatechnics" during World War II. Though the shift from polytechnic to monotechnic would seem to coincide with absolute monarchy, the Enlightenment, and the rise of manufacturing, Mumford is careful to point out that these ideas sometimes existed side by side and that one of the best examples of technocratic dictatorship was the pyramid age in Egypt.
54. Samuel W. Johnson, a professor at Yale University, was one of the most important early thinkers in what became agricultural science. Rossiter, *Emergence of Agricultural Science*, 149–71, passim.
55. Duncan takes up this subject in "English Capitalism in Its Light-Industrial Prime," in *Centrality of Agriculture*.

EPILOGUE

1. Kline does not calculate his or his family's time as a wage and might not figure the cost of his hay and other fodder to arrive at this sum.
2. Jennifer Vincent, *Michigan Farm News* (Aug. 30, 1999), at Web site www.michiganfarmbureau.com (June 22, 2001); Subbarat Muthukrishnan, "Wheat Crop Improvement by Genetic Engineering," Kansas State University Department of Biochemistry (ND), www.kswheat.com (July 23, 2001); Roxana Hegeman, writing for the Associated Press, "Kansas Wheat Harvest," web.northscape.com (July 23, 2001).
3. Census information from the National Agricultural Statistics Service of the United States Department of Agriculture, read from the Government Information Sharing Project of Oregon State University, govinfo.kerr.orst.edu (June 2001). There is a vigorous trade in horse-drawn implements and their parts among the Amish. A number of manufacturers make them new, including Pequea Machines of Gordonville, Pennsylvania, which builds an eighty-five-bushel-capacity manure spreader, the ideal size "for small farms or the horse farmer; the good news about Pequea is spreading almost as fast as the manure!" *Farming: People, Land, and Community* 1 (Spring 2001).
4. Nathan Weaver, "Why Grass Is Better," *Farming* 1 (Spring 2001), 10–14.
5. Dudley, *Debt and Dispossession*, 25–28.
6. Kline, *Great Possessions*, xx.
7. Ibid., 109.
8. For an interpretation of Dubos's work cast within environmentalism, see Donald Fleming, "Roots of the New Conservation Movement," *Perspectives in American History* 6 (1972), 34–39.
9. Dubos, *Wooing of Earth*, 5.

Bibliography

Manuscripts

Morris, James Pemberton. Diary, Jan. 1, 1823–Oct. 14, 1825. Joseph Downs Collection. Winterthur Library, Winterthur, Del.

Natural Science Manuscript Group. Manuscripts and Archives. Yale University.

Norton, John Pitkin. Papers. Manuscripts and Archives. Yale University.

Philadelphia Society for Promoting Agriculture. Records. Van Pelt Library, University of Pennsylvania.

Price, John. Diary, 1831–1847. The Historical Society of Pennsylvania.

Records of Ante-bellum Southern Plantations from the Revolution through the Civil War. Microform Collection. Yale University.

Gilliland, William. Papers, 1829–1867.

Law, Thomas Cassells. Papers, 1810–1865.

Means, Mary Hart, Papers, 1846–1865.

Miller-Furman-Dabbs Family Papers, 1751–1865.

Singleton Family Papers.

Ruffin, Edmund. Papers. Brock Collection. Huntington Library, San Marino, Calif.

Newspapers, Journals, and Serials

American Agriculturist (New York)

American Farmer (Baltimore)

American Republican (West-Chester, Pa.)

Berks and Schuylkill Journal (Reading, Pa.)

Carlisle Republican (Carlisle, Pa.)

Cultivator (Albany, N.Y.)

De Bow's Review (New Orleans, La.)

Farmers' Register (Shellbanks, Va.)

Farming: People, Land, and Community

Garden and Forest (Cambridge, Mass.)

Genesee Farmer (Rochester, N.Y.)

Memoirs of the Board of Agriculture of the State of New York
Memoirs of the Pennsylvania Agricultural Society
Memoirs of the Philadelphia Society for Promoting Agriculture
Memoirs of the Society of Virginia for Promoting Agriculture
New England Farmer (Boston)
New-York Tribune
Niles' Register (Malvern, Pa.)
Overland Monthly and Out West Magazine
Plough Boy (Albany, N.Y.)
The Plough, the Loom, and the Anvil (Philadelphia)
Soil of the South (Columbus, Ga.)
Southern Agriculturist, Horticulturist, and Register of Rural Affairs (Charleston, S.C.)
Southern Cultivator (Augusta, Ga.)
Southern Quarterly Review (Charleston, S.C.)
Transactions of the New York State Agricultural Society
United States Soil Conservation Service, USDA. Soil Survey

Agricultural Addresses

Abbot, Abiel. *An address delivered before the Essex Agricultural Society at the agricultural exhibition in Danvers, October 17, 1821*. Andover, Mass., 1822.

Address of a Convention of Delegates of the State of Pennsylvania, for the purpose of promoting the State Agricultural and Manufacturing Interests, at Harrisburg, June 27, 1827.

Allen, Lewis F. *Address Delivered before the New-York State Agricultural Society . . . 18th January, 1849*. Albany, N.Y.: Weed, Parsons and Company, 1849.

Biddle, Nicholas. *Address Delivered before the Philadelphia Society for Promoting Agriculture at its annual meeting on the Fifteenth of January, 1822*. Philadelphia, 1822.

Buel, Jesse. *An Address Delivered before the Berkshire Agricultural Society, October 5th, 1837*. Albany, N.Y., 1837.

Carey, Mathew. *Address to the Farmers of the United States, on the Ruinous Consequences of Their Vital Interests*. Philadelphia: M. Carey and Son, 1821.

Chipman, Samuel. *Address Delivered before the Ontario Agricultural Society at Its Sixth Annual Meeting*. Canandaigua, N.Y., 1824.

"Col. Chappell's Address, delivered on Saturday the 6th of July, to a large meeting of the Planters and other gentlemen of the neighborhood of Columbia." *Papers Published by Order of the Agricultural Society of South Carolina*. Columbia, S.C.: Telescope Press, 1818.

Colman, Henry. *An Address Delivered October 2, 1839, before the Middlesex Society of Husbandmen and Manufacturers*. Boston: Weeks and Company, 1839.

———. *On Labor: An Address at the Annual Cattle shows of the Worcester and the Hampshire, Hampden, and Franklin Agricultural Societies . . . October 1838.* Boston: Otis, Broaders, and Company, 1839.

Daingerfield, William A. "Prince George's County, Address of the Vice President." *American Farmer* 1 (July 16, 1819).

Eddy, Thomas. "Address." In *An Examination into the Expediency of Establishing a Board of Agriculture in the State of New York.* Brooklyn, 1819.

"Extracts from an Address, delivered before the Greenville Agricultural Society, August 1841." *Southern Agriculturist,* n.s., 2 (Jan. 1842).

Garnett, James Mercer. *An Address Delivered before the Virginia Agricultural Society . . . October 28, 1818.* Fredericksburg, Va.: Green and Harrow, 1818.

———. "Address." *American Farmer* 4 (May 3, 1822).

Genet, Edmond Charles. *Address on the Means of Opening New Sources of Wealth for the Northern States Delivered on the 19th of October, 1821, before the Agricultural Society of the County of Rensselaer.* Troy, N.Y.: Francis Adancourt, 1821.

Goodwin, Isaac. *Address Delivered before the Worcester Agricultural Society, October 13, 1824.* Worcester, Mass.: William Manning, [1824].

Greeley, Horace. *What the Sister Arts Teach as to Farming: An Address before the Indiana State Agricultural Society at Its Annual Fair.* New York: Fowlers and Wells, 1853.

Hammond, James Henry. "Anniversary Oration of the State Agricultural Society of South Carolina, Read before the Society on the 25th November, 1841." *Proceedings of the Agricultural Convention and of the Agricultural Society of South Carolina.* Columbia, S.C.: Summer and Carroll, 1846.

———. *An Address Delivered before the South-Carolina Institute at Its First Annual Fair, on the 20th November, 1849.* Charleston, S.C.: Walker and James, 1849.

Hilgard, Eugene W. *Address on Progressive Agriculture and Industrial Education.* Jackson, Miss.: Clarion Books, 1873.

Hopkins, Samuel M. *An Address to the Agricultural Society of the County of Genesee, Delivered at Batavia.* Batavia, N.Y.: Miller and Blodgett, 1819.

Keep, John. *An Address to the Agricultural and the Ladies' Manufacturing Societies of Cortland County, October 17, 1822.* Homer, N.Y., 1823.

Law, Thomas Cassells. *Report on the Best System of Farming for Improving Our Lands, Delivered to Hartsville Farmers' Club, June 8, 1871.*

Logan, George. *An Address on the Errors of Husbandry in the United States Delivered before the Philadelphia Society for Promoting Agriculture.* Philadelphia, 1818.

Lowell, John. "An Address Delivered before the Massachusetts Agricultural Society at the Brighton Cattle Show, October 13, 1818." *Massachusetts Agricultural Repository and Journal.* Vol. 5. Boston: Wells and Lilly, 1819.

Madison, James. *An Address Delivered before the Agricultural Society of Albemarle, on Tuesday, May 12, 1818.* Richmond, Va.: Shepherd and Pollard, 1818.

Marsh, George P. *Address Delivered before the Agricultural Society of Rutland County, September 30, 1847.* Rutland, Vt., 1848.

Mason, Cyrus. *The Oration on the Thirteenth Anniversary of the American Institute.* New York: D. Appleton and Company, 1840.

Morris, James Pemberton. *Address Delivered before the Agricultural Society of Bucks County.* Philadelphia: Clark and Raser, 1823.

Morton, W. S. "Address of W. S. Morton, President of the Agricultural Society of Cumberland." *Farmers' Register* 6 (1838): 535–38.

Norton, John Pitkin. *An Address before the Hartford County Agricultural Society, Delivered October 15, 1847.* Hartford, Conn., 1847.

———. *Address Delivered at the Annual Show of the N.Y. State Agricultural Society at Buffalo, September 6, 1848.* Albany, N.Y., 1848.

———. *Address before the Hampshire, Franklin, and Hampden Agricultural Society.* Northampton, Mass., October 1849.

———. *Address of Professor Norton Delivered before the Hampden County Agricultural Society.* Springfield, Mass., 1851.

Philadelphia Society for Promoting Agriculture. *Address to the Citizens of Pennsylvania on the Importance of a more Liberal Encouragement of Agriculture.* Philadelphia, 1818.

Rhett, Edmund. *Who Is the Producer? An Address Delivered before the Beaufort Agricultural Society, August 1840.* Charleston, S.C.: Hayden and Burke, 1840.

Schuyler, Philip. *Address Delivered before the Saratoga Society for Promoting Agriculture & Domestick Manufactures, at Ballston Spa, October 11, 1821.* Saratoga Springs, N.Y., 1821.

Simms, William Gilmore. *The Social Principle: The True Source of National Permanence, an Oration Delivered before the Erosophic Society of the University of Alabama . . . December 13, 1842.* Tuscaloosa, Ala.: The Society, 1843.

Spafford, Horatio Gates [Agricola, pseud.]. *Hints to Emigrants, on the Choice of Lands; Particularly Addressed to Farmers in the North-Eastern States.* Albany, N.Y.: J. Buel, 1817.

Taylor, William J. "Anniversary Address Delivered before the State Agricultural Society of South Carolina . . . 1843." *Proceedings of the Agricultural Convention of the State Agricultural Society of South Carolina.* Columbia, S.C.: Summer and Carroll, 1846.

Tilghman, William. *An Address Delivered before the Philadelphia Society for Promoting Agriculture . . . January 18, 1820.* Philadelphia: William Fry, 1820.

Van Bergen, A. *Address Delivered before the New York State Agricultural Society.* Albany, N.Y.: Charles Van Benthuysen, 1840.

Vaux, Roberts. *Address Delivered before the Philadelphia Society for Promoting*

Agriculture at its annual meeting on the Eighteenth of January, 1825. Philadelphia: Port Folio Office, 1825.

Whitaker, Daniel K. *The Claims of Agriculture to Be Regarded as a Distinct Science: An Address Delivered in Charleston before the Agricultural Society of South Carolina*. Charleston, S.C., 1833.

INTERVIEWS

Keilty, Michael. Farmer, Litchfield, Conn., at his farm. May 2000.

Kline, David. Farmer, Fredericksburg, Ohio, at his farm. June 2001.

Meinert, Richard. Cooperative Extension Educator with the University of Connecticut Cooperative Extension Service. Oct. 4, 1999 (by telephone), Dec. 7, 1999 (by telephone), May 2000.

Nieminen, Arnold O. Dairy Unit Manager, University of Connecticut, at the Durham Agricultural Fair, Durham, Conn. Sept. 24, 1999.

PRIMARY BOOKS AND ARTICLES

Adams, Daniel. *The Agricultural Reader, Designed for the Use of Schools*. Boston: Richardson and Lord, 1824.

Affleck, Thomas. *Southern Rural Almanac and Plantation and Garden Calendar for 1851*. Adams County, Miss., 1851.

Agricultural Society of South Carolina. *Original Communications Made to the Agricultural Society of South Carolina; and Extracts from Select Authors on Agriculture*. Charleston, S.C., 1824.

Allardice, Robert Barclay. *Agricultural Tour of the United States and Upper Canada*. London, 1842.

Allen, Lewis F. *Rural Architecture: Being a complete description of Farm Houses, Cottages, and Out Buildings*. New York: C. M. Saxton, 1853.

American Guano Company. *Report of Experiments with American Guano*. New York: W. H. Arthur and Company, 1859.

Amos, William. *Minutes in Agriculture and Planting*. Boston and London, 1804.

Annual Reports of the Commissioner of Patents

Armstrong, John. *A Practical Farmer, a Treatise on Agriculture; comprising a concise history of its origin and progress, the present Condition of the Art Abroad and at Home, and the Theory and Practice of Husbandry*. Albany, N.Y.: Jesse Buel, 1819.

Bailey, L. H. *The Harvest of the Year to the Tiller of the Soil*. New York: Macmillan, 1927.

Bard, Samuel. *A Guide for Young Shepherds; or, Facts and Observations on the Character and Value of Merino Sheep*. New York, 1811.

Beal, W. J. *Grasses of North America for Farmers and Students.* New York: H. Holt and Company, 1887.

Bennett, Hugh Hammond. *Soil Erosion a National Menace.* Circular no. 33, United States Department of Agriculture (April 1928).

Benton, C., and S. F. Barry. *A Statistical View of the Number of Sheep in the Several Towns and Counties. . . .* Cambridge, Mass.: Folsom, Wells, and Thurston, 1837.

Benton, Thomas Hart. *Mr. Benton's Speeches on the Public Lands, Delivered in the Senate of the United States at the First Session of the 20th Congress.* Washington, D.C., 1828.

Berry, Wendell. *The Unsettling of America: Culture and Agriculture.* San Francisco: Sierra Club Books, 1977.

Bidwell, Percy Wells, and John I. Falconer. *History of Agriculture in the Northern United States, 1620–1860.* Washington, D.C.: Carnegie Institution of Washington, 1925.

Birkbeck, Morris. *Letters from Illinois.* London: Taylor and Hessey, 1818.

Black, S. H. "An Essay on the Intrinsic Value of Arable Land." *American Farmer* 2 (April 7, 1820).

Blake, John L. *The Farmer's Every-Day Book.* Auburn, N.Y.: Derby, Miller and Company, 1850.

Bordley, John Beale. *A Summary View of the Courses of Crops in the Husbandry of England and Maryland; with a Comparison of Their Products and a System of Improved Courses, proposed for Farms in America.* Philadelphia, 1784.

Bowen, Eli. *The Pictorial Sketch-Book of Pennsylvania; or, Its Scenery, Internal Improvements, Resources, and Agriculture.* Philadelphia: Willis P. Hazard, 1852.

Buckner, Richard. "Rural Economy—No. III." *American Farmer* 3 (April 6, 1821).

Buel, Jesse. *State of New-York Agricultural Almanac for the Year of Our Lord 1823.* Albany, N.Y.: Daniel Steele and Son, 1823.

———. *The Farmer's Companion; or, Essays on the Principle and Practice of American Husbandry.* 1839. 6th rev. ed., New York: Harper and Brothers, 1847.

Butler, Frederick. *The Farmer's Manual: Being a Plain Practical Treatise on the Art of Husbandry Designed to Promote an Acquaintance with the Modern Improvements in Agriculture.* Hartford, Conn.: Samuel G. Goodrich, 1819.

Carey, Henry C. *The Harmony of Interests: Agricultural, Manufacturing, and Commercial.* Philadelphia: Henry Carey Baird, 1872. Repr., Henry C. Carey, ed. *Miscellaneous Works.* Philadelphia, 1883.

Carey, Mathew. *The New Olive Branch; or, An Attempt to Establish an Identity of Interest between Agriculture, Manufacturing, and Commerce.* Philadelphia: M. Carey and Son, 1820.

———. *(Private) Sir, Although it is, in general, improper for a man to write or speak of himself, yet there are occasions in which it is allowable.* Broadside from the American Antiquarian Society, 1826.

———. Pennsylvania Society for the Promotion of Manufacturers and the Mechanic Arts. "At a very numerous and respectable meeting of Farmers and Manufacturers. . . ." [Philadelphia], May 31, 1827. American Antiquarian Society.

[Carey?]. *To the Honourable [Members of] the Senate and House of Representatives of the United States, in Congress assembled, The Memorial of the Subscribers, Citizens of the City and County of Philadelphia.* 1827.

Carman, Harry J., ed. *Jesse Buel, Agricultural Reformer: Selections from His Writings.* New York: Columbia University Press, 1947.

Cato, Marcus Porcius. *Cato the Censor on Farming.* New York: Columbia University Press, 1933.

Census of 1820. Washington, D.C.: Gales and Seaton, 1821. Repr., New York: Norman Ross Publishing, 1990.

Chaderton, William. *Classical, English, and Agricultural Institute at Bolton Farm, Bucks County, Pa.* Philadelphia: Philadelphia Society for Promoting Agriculture, c. 1830.

Clay, Henry. *Mr. Clay's Speech upon the Tariff, or The "American System," So Called, or the Anglican System, in Fact. . . .* Richmond, Va., 1827.

Cobbett, William. *A Year's Residence in the United States of America.* 3rd ed. London, 1822.

Commemorative Biographical Record of Central Pennsylvania. 1898. Typescript photocopy in the Pennsylvania State University Library.

Compendium of the Enumeration of the Inhabitants and Statistics of the United States as obtained at the Department of State, from the Returns of the Sixth Census. Washington, D.C.: Thomas Allen, 1841.

Constitution of the Merino Society of the Middle States of North America. Philadelphia, 1811.

Cooper, Ellwood. *Forest Culture and Eucalyptus Trees.* San Francisco: Cubery and Company, 1876.

Cooper, James Fenimore. *The Pioneers.* 1823. Repr., New York: Penguin Books, 1988.

Cooper, Susan Fenimore. *Rural Hours.* 3rd ed. New York: G. P. Putnam, 1851.

Copeland, Robert Morris. *Country Life: A Handbook of Agriculture, Horticulture, and Landscape Gardening.* 5th rev. ed. Boston: Dinsmoor and Company, 1866.

Coxe, Tench. *An Addition, of December 1818, to the Memoir, of February and August 1817, on the subject of Cotton Culture, the Cotton Commerce, and the Cotton Manufacture of the United States.* Philadelphia, 1818.

Darby, William. *The Emigrant's Guide to the Western and Southwestern States and Territories.* New York: Kirk and Mercein, 1818.

———. *A Tour from the City of New York to Detroit, in the Michigan Territory,*

Made between the 2d of May and the 22d of September, 1818. New York, 1819. Repr., Chicago: Quadrangle Books, 1962.

Darwin, Erasmus. *Phytologia; or, The Philosophy of Agriculture and Gardening*. Dublin: P. Byrne, 1800.

Davis, Alexander Jackson. *Rural Residences, etc.: Consisting of Designs, Original and Selected, for Cottages, Farm-Houses, Villas, and Village Churches*. New York, 1837.

Davis, David Brion, ed. *Antebellum American Culture: An Interpretive Anthology*. Lexington, Mass.: D. C. Heath and Company, 1979.

Demaree, Albert Lowther, ed. *The American Agricultural Press, 1819–1860*. New York: Columbia University Press, 1941.

Destutt de Tracy, Antoine-Louis-Claude. *A treatise on Political Economy*, trans. Thomas Jefferson. Georgetown, D.C.: J. Milligan, 1817.

Digest of Accounts of Manufacturing Establishments in the United States and of their Manufactures. Made under the direction of the Secretary of State, in pursuance of a Resolution of Congress, of 30th March, 1822. Washington, D.C.: Gales and Seaton, 1823.

Downing, Andrew Jackson. *Cottage Residences; or, A series of designs for Rural Cottages and Cottage-Villas and their Gardens and Grounds adapted to North America*. New York and London: Wiley and Putnam, 1842.

———. *A treatise on the theory and practice of landscape gardening, adapted to North America; with a view to the improvement of country residences*. 4th rev. ed. New York: G. P. Putnam, 1853.

———. *Rural Essays*. New York: R. Worthington, 1881.

Drayton, John. *A View of South-Carolina*. Charleston, S.C.: W. P. Young, 1802.

Drown, William, and Solomon Drown. *Compendium of Agriculture; or, The Farmer's Guide, in the most essential parts of Husbandry and Gardening*. Providence, R.I.: Field and Maxcy, 1824.

Dudley, Kathryn Marie. *Debt and Dispossession: Farm Loss in America's Heartland*. Chicago: University of Chicago Press, 2000.

Eaton, Amos. *Geological and Agricultural Survey of Rensselaer County in the State of New York*. Albany, N.Y., 1822.

Eliot, Jared. *Essays upon Field Husbandry in New England, and Other Papers, 1748–1762*, ed. Harry J. Carman and Rexford G. Tugwell. New York: Columbia University Press, 1934.

Ellsworth, T. H. L. *The Improvements in Agriculture and the Arts of the United States . . . a Report to the Congress of the United States*. [Washington, D.C.], 1844.

Estwick, Evans. *A Pedestrious Tour of Four Thousand Miles through the Western States and Territories*. Concord, N.H.: Joseph Spear, 1819.

Everett, Alexander H. *New Ideas on Population: With Remarks on the Theories of Malthus and Godwin*. Boston: Oliver Everett, 1823.

Faux, W. *Memorable Days in America: Being a Journal of a Tour to the United States*. London: W. Simpkin and R. Marshall, 1823.

Fearon, Henry Bradshaw. *Sketches of America: A Narrative Journey of Five Thousand Miles through the Eastern and Western States of America*. London: Longman, 1818.

Federal Writers' Program of the Works Projects Administration. *Pennsylvania: A Guide to the Keystone State*. New York: Oxford University Press, 1940.

Fessenden, Thomas Green. *The Complete Farmer and Rural Economist*. 2nd ed. Boston, 1835.

———. *The Farmer's Road to Wealth, Containing a calendar of work for the Farm*. Boston, [1834].

Flint, James. *Letters from America*. Edinburgh: W. and C. Tait, 1822.

Fowler, John. *Journal of a Tour in the State of New York, in the year 1830; with Remarks on Agriculture*. London: Whittaker, Treacher, and Arnot, 1831.

Garnett, James M. "Defects in Agriculture." *Memoirs of the Society of Virginia for Promoting Agriculture*. Richmond, Va.: Shepherd and Pollard, 1818.

General Convention of Agriculturists and Manufacturers and Others Friendly to the Encouragement and Support of the Domestic Industry of the United States, Monday, July 30, 1827. Pamphlet in the collection of New York University.

Goodrich, Samuel G. *Recollections of a Lifetime; or, Men and Things I Have Seen*. Vol. 1. New York: Miller, Orton and Mulligan, 1856.

Gray, Alonzo. *Elements of Scientific and Practical Agriculture*. New York: Dayton and Newman, 1842.

Greeley, Horace. *An Overland Journey from New York to San Francisco in the Summer of 1859*. New York: C. M. Saxton Baker, 1860.

———. *Mr. Greeley's Letters from Texas and the lower Mississippi*. New York: Tribune, 1871.

———. *What I Know about Farming*. New York: Tribune, 1871.

Grisenthwaite, W. "A New Theory of Agriculture." *American Farmer* 3 (June 1, 1821).

Hammond, James Henry. "Southern Industry." A series of articles compiled from *De Bow's Review* and published in *The Southern States, Embracing a Series of Papers Condensed from the Earlier Volumes of "De Bow's Review."* Washington, D.C., 1856.

———. *Selections from the Letters and Speeches of the Hon. James Henry Hammond of South Carolina*. New York: John F. Trow, 1866.

———. *Secret and Sacred: The Diaries of James Henry Hammond, a Southern Slaveholder*, ed., Carol Bleser. New York: Oxford University Press, 1988.

Hammond, J. H. *The Farmer's and Mechanic's Practical Architect; and Guide in Rural Economy*. Boston: John P. Jewett and Company, 1858.

Harper, L. *Preliminary Report on the Geology and Agriculture of the State of Mississippi*. Jackson, Miss., 1857.

Harris, David Golightly. *Piedmont Farmer: The Journals of David Golightly Harris, 1855–1870*, ed. Philip N. Racine. Knoxville: University of Tennessee Press, 1990.

Hilgard, Eugene W. *Report of the Geological and Agricultural Survey of the State of Mississippi*. Jackson, Miss.: Mississippian Steam Power Press, 1858.

Hints for American Husbandmen with Communications to the Pennsylvania Agricultural Society. Philadelphia: Clark and Raser, 1827.

Howe, Henry. *Adventures and Achievements of Americans: A series of narratives illustrating their heroism, self-reliance, genius, and enterprise.* New York: G. F. Tuttle, 1859.

Howitt, E. *Selections from Letters Written during a Tour through the United States, in the Summer and Autumn of 1819.* Nottingham, U.K.: J. Dunn, n.d.

Humphreys, David. *A Discourse on the Agriculture of the State of Connecticut and the Means of Making It More Beneficial to the State.* New Haven, Conn., 1816.

Illustrated Souvenir History of Philipsburg, Pennsylvania. Williamsport, Pa.: Grit Publishing Company, 1909.

Jay, John. *A Statistical View of American Agriculture, its home resources and foreign markets, with suggestions for the schedules of the federal census in 1860.* New York: D. Appleton and Company, 1859.

Jeffreys, George W. [Agricola, pseud.]. *A Series of Essays on Agriculture & Rural Affairs.* Raleigh, N.C.: Gales, 1819.

Johnson, S. W. *Rural Economy: Containing a Treatise on Pisé Building; on Building in General; on the Culture of the Vine; and on Turnpike Roads.* New York, 1806.

Journal of a Convention of Delegates of the State of Pennsylvania, Held for the Promotion of the State Agricultural and Manufacturing Interests . . . 27th of June, A.D. 1827. [Harrisburg, Pa.], 1827.

Laws and Regulations of the Massachusetts Society for Promoting Agriculture . . . by the Trustees. Boston, 1793.

Lee, Daniel. "Philosophy of Tillage." *Transactions of the N.Y. State Agricultural Society . . . 1848.* New York: Weed, Parsons and Company, 1848.

Liebig, Justus von. *Die organische Chemie in ihrer Anwendung auf Agricultur und Physiologie.* Braunschweig: F. Vieweg, 1842.

———. *Chemistry in Its Application to Agriculture and Physiology*, edited from the manuscript of the author by Lyon Playfair. Philadelphia: James Campbell, 1843.

———. *The Relations of Chemistry to Agriculture and the Agricultural Experiments of Mr. J. B. Lawes*, trans. Samuel Johnson. Albany, N.Y.: Luther Tucker, 1855.

Linn, John Blair. *History of Centre and Clinton Counties, Pennsylvania.* 1883. Repr., Centre County Historical Society Bicentennial Edition, 1975.

Livingston, Robert R. *Essay on Sheep; Their Varieties—Account of the Merinoes of Spain, France, &c.* New York: T. and J. Swords, 1809.

Logan, George. *A Letter to the Citizens of Pennsylvania on the Necessity of Promoting Agriculture, Manufactures, and the Useful Arts.* Lancaster, Pa., March 14, 1800.

Lorain, John. *Hints to Emigrants; or, A Comparative Estimate of the Advantages of Pennsylvania and of the Western Territory.* Philadelphia: Little and Henry, 1819.

———. *Nature and Reason Harmonized in the Practice of Husbandry.* Philadelphia: H. C. Carey and I. Lea, 1825.

Marsh, Caroline Crane, ed. *The Life and Letters of George Perkins Marsh.* New York: Scribner's, 1888.

Marsh, George Perkins. *Man and Nature; or, Physical Geography as Modified by Human Action*, ed., David Lowenthal. 1864. Repr., Cambridge, Mass.: Harvard University Press, 1965.

Mead, Elwood. "Government Aid and Direction in Land Settlement." *American Economic Review* 8 (March 1918): 72–98.

———. *Irrigation Institutions, a Discussion of the Economic and Legal Questions Created by the Growth of Irrigated Agriculture in the West.* New York: The Macmillan Company, 1903.

Meeker, Nathan Cook. *Life in the West; or, Stories of the Mississippi Valley.* New York: S. R. Wells, 1868.

Melish, John. *Map of Pennsylvania.* [Philadelphia?]: State of Pennsylvania, 1822.

Memorial of the Agricultural Society of South Carolina Adverse to an Increase of Duties on Coarse Woollens and Other Imports. 20th Cong., 1st sess., Jan. 9, 1828. S. Doc. 26. Washington, D.C.: Duff Green, 1828.

Memorial of the Ontario Agricultural Society, of the State of New York, Praying Further Protecting Duties on Certain Domestic Manufacturers. 20th Cong., 1st sess., Jan. 14, 1828. H. Doc. 63. Washington, D.C.: Gales and Seaton, 1828.

Merino Sheep: Authentic Documents, relative to the quality and breed of Merino Sheep . . . , New York, 1810. Broadside from the American Antiquarian Society, Worcester, Mass.

Michaux, F. A. *Travels to the Westward of the Allegheny Mountains in the States of the Ohio, Kentucky, and Tennessee . . . in the Year X, 1802*, trans. B. Lambert. London, 1805.

Mills, Robert. *Mills' Atlas of the State of South Carolina.* 1825. Repr., Easley, S.C.: Southern Historical Press, 1980.

———. *Statistics of South Carolina.* Charleston, S.C.: Hurlbut and Lloyd, 1826.

National Atlas of the United States of America. Washington, D.C.: United States Department of the Interior, Geological Survey, 1970.

Nearing, Helen, and Scott Nearing. *Living the Good Life.* 1954. Repr., New York: Schocken Books, 1989.

Nicholson, John. *The Farmer's Assistant: Being a Digest of All That Relates to Agriculture and the Conduct of Rural Affairs.* Lancaster, Pa.: Benjamin Warner, 1819.

Odum, Howard W. *Southern Regions of the United States.* Chapel Hill: University of North Carolina Press, 1936.

Olmsted, Frederick Law. *Walks and Talks of an American Farmer.* New York: G. P. Putnam, 1852.

Papers Published by Order of the Agricultural Society of South Carolina. Columbia, S.C.: Telescope Press, 1818.

Parkinson, Richard. *Of Turnip and Pea Fallows, with a design of a Rotation of Crops, Recommended to the Farmers and Planters of the United States of America.* Washington, D.C.: 1801.

Pearson, John. *Notes during a journey in 1821 in the United States of America: from Philadelphia to the neighborhood of Lake Erie; through Lancaster, Harrisburgh, Carlisle & Pittsburgh and back to Philadelphia . . . in search of a settlement.* London: W. and S. Couchmann, 1822.

Pennsylvania Agricultural & Manufacturing Company. 1813.

Pennsylvania Agricultural Society. American Antiquarian Society, n.d.

Pinchot, Gifford. *The Fight for Conservation.* New York: Doubleday, 1910.

———. *Breaking New Ground.* New York: Harcourt, Brace, and Company, 1947. Repr., Washington, D.C.: Island Press, 1998.

Pitkin, Timothy. *A Statistical View of the Commerce of the United States of America: Its Connection with Agriculture and Manufactures.* Hartford, Conn., 1816.

Preston, John. *A Statistical Report of the County of Albany, for the Year 1820.* Albany, N.Y.: Packard and Van Benthuysen, 1823.

Proceedings of the Pennsylvania Agricultural Society with the Address of their President Jonathan Roberts, Esq. Philadelphia: Clark and Raser, 1823.

Register of Debates in Congress

Report of the Secretary of the Treasury communicating Statistical information in relation to the condition of the agriculture, manufactures, domestic trade, currency, and banks, of the United States. . . . 28th Cong., 2nd sess., Jan. 6, 1845. S. Doc. 21.

Resolutions and Remonstrance of the Agricultural Society of St. Andrew's Parish, S.C. against any Further Increase of the Duties on Imports, &c. 20th Cong., 1st sess. Dec. 14, 1827. H. Doc. 13. Washington, D.C.: Gales and Seaton, 1827.

Resolutions and Remonstrance of the Agricultural Society of St. John's, Colleton, S.C. against any Further Increase of the Duties on Imports, &c. 20th Cong., 1st sess., Dec. 14, 1827. H. Doc 14. Washington, D.C.: Gales and Seaton, 1827.

Review and Abstract of the County Reports to the Board of Agriculture. York, U.K.: Thomas Wilson and Sons, 1818.

Robinson, Solon. *Guano: A treatise of practical information for farmers; containing plain directions how to apply Peruvian guano to the various crops and soils of America.* New York, 1852.

———. *Solon Robinson, Pioneer and Agriculturist: Selected Writings*, ed. Herbert Anthony Kellar. 2 vols. Indianapolis: Indiana Historical Bureau, 1936.

———., ed. *Facts for Farmers; Also for the Family Circle. A Variety of Rich Materials for All Land-Owners about Domestic Animals and Domestic Economy.* New York, 1869.

Rodale, J. I., ed. *How to Grow Vegetables and Fruits by the Organic Method.* Emmaus, Pa.: Rodale Press, 1961.

Ruffin, Edmund. *An Essay on Calcareous Manures.* Shellbanks, Va.: Farmers' Register, 1835.

———. *Report of the Commencement and Progress of the Agricultural Survey of South Carolina, for 1843.* Columbia, S.C.: State Printer, 1843.

———. *Agricultural, Geological, and Descriptive Sketches of Lower North Carolina, and the Similar Adjacent Lands.* Raleigh, N.C., 1861.

———. *Agriculture, Geology, and Society in Antebellum South Carolina: The Private Diary of Edmund Ruffin*, ed. William Mathew. Athens: University of Georgia Press, 1992.

Rush, Benjamin. *Essays, Literary, Moral, and Philosophical.* Philadelphia: Bradford, 1806.

Say, Jean-Baptiste. *A Treatise on Political Economy; or, The Production, Distribution, and Consumption of Wealth.* London, 1821.

Schweinitz, Lewis David von. *Journey . . . to Goshen, Bartholomew County, in 1831.* Indianapolis: Indiana Historical Society Publications 8 (1927).

Seventh Census of the United States: 1850. Washington, D.C.: Robert Armstrong, 1853.

Shaler, Nathaniel Southgate. *Man and the Earth.* New York: Duffield and Company, 1905. Repr., 1910.

Simms, William Gilmore. *Richard Hurdis: A Tale of Alabama.* New York: Redfield, 1855.

Sinclair, John. *An Account of the Systems of Husbandry of Scotland.* N.p., 1813.

———. *The Code of Agriculture; Including Observations on Gardens, Orchards, Woods, and Plantations . . . First American Edition with Notes.* Hartford, Conn.: Hudson and Company, 1818.

Spafford, Horatio Gates. *A Gazetteer of the State of New York.* Albany, N.Y.: H. C. Southwick, 1813.

Stirling, James. *Letters from the Slave States.* London: John W. Parker, 1857.

Strickland, William. *Observations on the Agriculture of the United States of America.* London: Bulmer and Co., 1801. Published as part of Reverend J. E. Strickland, ed., *Journal of a Tour in the United States of America, 1794–1795, by William Strickland.* New York: New-York Historical Society, 1971.

Surveyor General of the Commonwealth of Pennsylvania. Annual Report.

Tatham, William. *Communications concerning the Agriculture and Commerce of the United States of America: Being an auxiliary to a report made by William Strickland*. London: J. Ridgway, 1800.

Taylor, Jefferys. *The Farm; or, A New and Entertaining Account of Rural Scenes and Pursuits with the Toils and Pleasures, and Productions of Farming: For Young Readers in the Town and Country*. Boston: Carter, Hendee, and Company, 1834.

Taylor, John. *Arator: Being A Series of Agricultural Essays, Practical and Political, in Sixty-four Numbers*, ed. M. E. Bradford. Indianapolis: Liberty Fund, 1977.

Thomson, William. *A Tradesman's Travels in the United States and Canada, in the Years 1840, 41, & 42*. Edinburgh: Oliver and Boyd, 1842.

Thoreau, Henry David. *Walden*. 1854. Repr., New York: Penguin American Library, 1983.

———. *The Maine Woods*. 1864. Repr., New York: Penguin Books, 1988.

Tibbits, George. *Essay on the Expediency and Practicability of Improving or Creating Home Markets for the Sale of Agricultural Productions and Raw Materials by the Introduction and Growth of Artisans and Manufacturers*. Philadelphia: Clark and Raser, 1829.

To the Honorable the Senate and House of Representatives of the United States in Congress Assembled, The Memorial of the Subscribers, Farmers of the State of Pennsylvania. Dec. 15, 1823. American Antiquarian Society.

Toulmin, Harry. *The Western Country in 1793: Reports on Kentucky and Virginia*, ed. Marion Tinling and Godfrey Davies. San Marino, Calif.: Henry E. Huntington Library and Art Gallery, 1948.

Tucker, Luther. *The Cultivator Almanac for the Year 1848*. Clyde, N.Y.: Cultivator, 1848.

Tull, Jethro. *The horse-hoeing husbandry; or, An essay on the principles of tillage and vegetation*. London: Printed for the author, 1733.

Tuomey, M. *Report on the Geology of South Carolina*. Columbia, S.C.: A. S. Johnston, 1848.

Turner, J. A., ed. *The Cotton Planter's Manual: Being a Compilation of Facts from the Best Authorities on the Culture of Cotton*. New York: C. M. Saxton, 1857.

United States Bureau of Animal Industry. *Special Report on the Sheep Industry*. Washington, D.C., 1892.

United States Soil Conservation Service, USDA. *Soil Survey—Newberry County, South Carolina*. Series 1956, no. 10. Washington, D.C.: Government Printing Office, 1960.

———. *Soil Survey—Chester and Delaware Counties, Pennsylvania*. Washington, D.C.: Government Printing Office, 1963.

Varlo, Charles. *The Essence of Agriculture: Being a Regular System of Husbandry, through all its Branches, Suited to the Climate and Lands of Ireland . . . with the Author's twelve Months Tour thro' America*. London, 1786.

Warden, D. B. *A Statistical, Political, and Historical Account of the United States of North America*. 3 vols. Philadelphia: Thomas Wardel, 1819.

Washington, George. *Letters from His Excellency George Washington to Arthur Young, ESQ., F.R.S., and Sir John Sinclair, BART., M.P.: Containing an Account of His Husbandry with His Opinions on Various Questions in Agriculture*. Alexandria, Va.: Cottom and Stewart, 1803.

Watson, Elkanah. *History of the Rise, Progress, and Existing Condition of the Western Canals in the State of New York . . . Together with the Rise, Progress, and Existing State of Modern Agricultural Societies on the Berkshire System*. Albany, N.Y.: D. Steele, 1820.

Webster, Daniel. *Speech of Daniel Webster in Reply to Mr. Hayne, of South Carolina: The Resolution of Mr. Foot of Connecticut Relative to the Public Lands . . . January 26, 1830*. Washington, D.C.: Gales and Seaton, 1830.

Wiggins, Francis. *The American Farmer's Instructor*. Philadelphia: O. Rogers, 1840.

William and Eliza; or, The Visit. New York: Solomon King, [1828?].

Young, Arthur. *A Six Weeks Tour, through the Southern Counties of England and Wales*. London, 1769.

———. *A Course of Experimental Agriculture*. London, 1770.

———. *Political Arithmetic: Containing Observations on the Present State of Great Britain*. London, 1774.

———. *Travels in France during the Years 1787, 1788, 1789*. 1792. Repr., Cambridge: Cambridge University Press, 1929.

———. *General Report on Enclosures, Drawn up by Order of the Board of Agriculture*. London, 1808.

SECONDARY BOOKS AND ARTICLES

Adams, Robert McC. *Heartland of Cities: Surveys of Ancient Settlement and Land Use on the Central Floodplain of the Euphrates*. Chicago: University of Chicago Press, 1981.

Allen, Robert C., and Cormac O. Grada. "On the Road Again with Arthur Young: English, Irish, and French Agriculture during the Industrial Revolution." *Journal of Economic History* 48 (March 1988): 93–116.

Allmendinger, David F. *Ruffin: Family and Reform in the Old South*. New York: Oxford University Press, 1990.

Appleby, Joyce. "The 'Agrarian Myth' in the Early Republic." In *Liberalism and Republicanism in the Historical Imagination*. Cambridge, Mass.: Harvard University Press, 1992.

Atack, Jeremy, and Fred Bateman. *To Their Own Soil: Agriculture in the Antebellum North*. Ames: Iowa State University Press, 1987.

Barger, Harold, and Hans H. Landsberg. *American Agriculture: A Study of Output, Employment, and Productivity*. New York: National Bureau of Economic Research, 1942.

Barron, Hal S. *Those Who Stayed Behind: Rural Society in Nineteenth-Century New England*. New York: Cambridge University Press, 1984.

Barton, Gregory. "Sir Albert Howard and the Forestry Roots of the Organic Farming Movement." *Agricultural History* 75 (Spring 2001): 168–87.

Beer, Max. *An Inquiry into Physiocracy*. London: Allen and Unwin, 1939.

Boyd, David. *A History: Greeley and the Union Colony of Colorado*. Greeley, Colo.: Greeley Tribune Press, 1890.

Brady, Nyle C. *The Nature and Properties of Soils*. 10th ed. New York: Macmillan, 1990.

Brown, Jonathan, and H. A. Beecham. "Arable Farming." In *The Agrarian History of England and Wales. Vol. VI, 1750–1850*, ed. G. E. Mingay. Cambridge: Cambridge University Press, 1989.

Carman, Harry J. "Jesse Buel, Albany County Agriculturalist." *New York History* 14 (July 1933).

Cauthen, Charles Edward. "The Secession Movement before 1860." Reprinted in *Perspectives in South Carolina History*, ed. Ernest M. Lander, Jr., and Robert K. Ackerman. Columbia: University of South Carolina Press, 1973.

Cayton, Andrew R. L. *The Frontier Republic: Ideology and Politics in the Ohio Country, 1780–1825*. Kent, Ohio: Kent State University Press, 1986.

Chambers, J. D., and G. E. Mingay. *The Agricultural Revolution, 1750–1880*. New York: Schocken Books, 1966.

Chaplin, Joyce. *An Anxious Pursuit: Agricultural Innovation and Modernity in the Lower South, 1730–1815*. Chapel Hill: University of North Carolina Press, 1993.

Clark, Christopher. *The Roots of Rural Capitalism: Western Massachusetts, 1780–1860*. Ithaca, N.Y.: Cornell University Press, 1990.

Clark, Thomas D., and John D. W. Guice. *Frontiers in Conflict: The Old Southwest, 1795–1830*. Albuquerque: University of New Mexico Press, 1989.

Coclanis, Peter A. *The Shadow of a Dream: Economic Life and Death in the South Carolina Low Country, 1670 to 1920*. New York: Oxford University Press, 1989.

Cole, H. H., and W. N. Garrett. *Animal Agriculture: The Biology, Husbandry, and Use of Domestic Animals*. 2nd ed. San Francisco: W. H. Freeman and Company, 1980.

Conkin, Paul K. "The Vision of Elwood Mead," *Agricultural History* 34 (April 1960): 88–97.

Conzen, Michael P., ed. *The Making of the American Landscape*. New York: Routledge, 1994.

Cooper, William J., Jr. *The South and the Politics of Slavery, 1828–1856*. Baton Rouge: Louisiana State University Press, 1978.

Craven, Avery. *Soil Exhaustion as a Factor in the Agricultural History of Virginia and Maryland, 1606–1860*. Urbana: University of Illinois, 1926.

———. *Edmund Ruffin, Southerner; A Study in Secession*. Hamden, Conn.: Archon Books, [c. 1932].

———. "John Taylor and Southern Agriculture." *Journal of Southern History* 4 (May 1938): 137–47.

Dangerfield, George. *The Era of Good Feelings*. New York: Harcourt, Brace and Company, 1952.

Danhof, Clarence H. *Change in Agriculture: The Northern States, 1820–1870*. Cambridge, Mass.: Harvard University Press, 1969.

Deane, Phyllis, and A. W. Cole. *British Economic Growth, 1688–1959*. Cambridge: Cambridge University Press, 1962.

Dorman, Robert L. *A Word for Nature: Four Pioneering Environmental Advocates, 1845–1913*. Chapel Hill and London: University of North Carolina Press, 1998.

Dubos, René. *The Wooing of Earth*. New York: Scribner, 1980.

Duncan, Colin A. M. *The Centrality of Agriculture: Between Humankind and the Rest of Nature*. Montreal: McGill-Queen's University Press, 1996.

Earle, Carville. "Myth of the Southern Soil Miner." In *The Ends of the Earth: Perspectives on Modern Environmental History*, ed. Donald Worster. New York: Cambridge University Press, 1988.

Eisenberg, Evan. *The Ecology of Eden*. New York: Knopf, 1998.

Ensminger, Robert F. *The Pennsylvania Barn: Its Origin, Evolution, and Distribution in North America*. Baltimore: Johns Hopkins University Press, 1992.

Faragher, John Mack. *Sugar Creek: Life on the Illinois Prairie*. New Haven, Conn.: Yale University Press, 1986.

Faust, Drew Gilpin. "The Rhetoric and Ritual of Agriculture in Antebellum South Carolina." *Journal of Southern History* 45 (Nov. 1979): 541–68.

———. *James Henry Hammond and the Old South: A Design for Mastery*. Baton Rouge: Louisiana State University Press, 1982.

Feller, Daniel. *The Public Lands in Jacksonian Politics*. Madison: University of Wisconsin Press, 1984.

———. *The Jacksonian Promise: America, 1815–1840*. Baltimore: Johns Hopkins University Press, 1995.

Fletcher, Stevenson Whitcomb. *Pennsylvania Agriculture and Country Life, 1640–1840*. Harrisburg: Pennsylvania Historical and Museum Commission, 1950.

Freehling, William W. *Prelude to Civil War: The Nullification Controversy in South Carolina*. New York: Harper and Row, 1965.

Gates, Paul W. *The Farmer's Age: Agriculture, 1815–1860*. New York: Harper Torchbooks, 1960.

Genovese, Eugene D. *The Political Economy of Slavery: Studies in the Economy and Society of the Slave South*. New York: Vintage, 1965.

Golley, Frank B. *A Primer for Environmental Literacy*. New Haven, Conn.: Yale University Press, 1998.

Grammer, John M. *Pastoral and Politics in the Old South*. Baton Rouge: Louisiana State University Press, 1996.

Grove, Richard. *Green Imperialism: Colonial Expansion, Tropical Island Edens, and the Origins of Environmentalism, 1600–1860*. Cambridge: Cambridge University Press, 1995.

Hahn, Stephen. *The Roots of Southern Populism: Yeoman Farmers and the Transformation of the Georgia Upcountry*. New York: Oxford University Press, 1983.

Hall, Arthur R. *The Story of Soil Conservation in the South Carolina Piedmont, 1800–1860*. United States Department of Agriculture Miscellaneous Publication no. 407. Washington, D.C., Nov. 1940.

Hays, Samuel P. *Conservation and the Gospel of Efficiency*. Cambridge, Mass.: Harvard University Press, 1959. Repr., Pittsburgh: University of Pittsburgh Press, 1999.

Henretta, James A. "Families and Farms: Mentalité in Pre-industrial America." *William and Mary Quarterly* 35 (Jan. 1978): 3–32.

Herman, Bernard L. "The Model Farmer and the Organization of the Countryside." In *Everyday Life in the Early Republic*, ed. Catherine E. Hutchins. Winterthur, Del.: Henry Francis du Pont Winterthur Museum, 1994.

Herndon, Melvin G. "Agriculture in America in the 1790s: An Englishman's View." *Agricultural History* 49 (July 1975): 505–16.

Hietala, Thomas R. *Manifest Design: Anxious Aggrandizement in Late Jacksonian America*. Ithaca, N.Y.: Cornell University Press, 1985.

Howe, Daniel Walker. *The Political Culture of the American Whigs*. Chicago: University of Chicago Press, 1979.

Hudson, Norman. *Soil Conservation*. Yorkshire, U.K..: B. T. Batsford Limited, 1971.

Hunter, Brooke. "Grist for the Mill: The Refinement of an American Industry." Unpublished manuscript, 2000.

Hurt, R. Douglas. *Indian Agriculture in America*. Lawrence: University Press of Kansas, 1987.

Hyams, Edward. *Soil and Civilization*. London: Thames and Hudson, 1952.

Jacobsen, Thorkild. *Salinity and Irrigation Agriculture in Antiquity: Diyala Basin Archaeological Projects: Report on Essential Results, 1957–58*. Bibliotheca Mesopotamica, vol. 14. Malibu, Calif.: Undena Publications, 1982.

Jones, E. L. "Creative Disruptions in American Agriculture, 1620–1820." *Agricultural History* 48 (1974): 510–28.

Jordan, Terry G., and Matti Kaups. *The American Backwoods Frontier: An Ethnic and Ecological Interpretation*. Baltimore: Johns Hopkins University Press, 1989.

Judd, Richard. *Common Lands, Common People: The Origins of Conservation in Northern New England*. Cambridge, Mass.: Harvard University Press, 1997.

Kerridge, Eric. *The Agricultural Revolution*. London: Allen and Unwin, 1967.

———. *The Farmers of Old England*. London: Allen and Unwin, 1973.

Kirby, Jack Temple. *Poquosin: A Study of Rural Landscape and Society*. Chapel Hill: University of North Carolina Press, 1995.

———, ed. *Nature's Management: Writings on Landscape and Reform, 1822–1859*. Athens: University of Georgia Press, 2000.

Klein, Philip Shriver. *Pennsylvania Politics, 1817–1832, a Game without Rules*. Philadelphia: Historical Society of Pennsylvania, 1940.

Klein, Rachel N. *The Unification of South Carolina: The Rise of the Planter Class in the South Carolina Backcountry, 1760–1808*. Chapel Hill: University of North Carolina Press, 1990.

Kline, David. *Great Possessions: An Amish Farmer's Journal*. Foreword by Wendell Berry. San Francisco: North Point Press, 1990.

Knobloch, Frieda. *The Culture of Wilderness: Agriculture as Colonization in the American West*. Chapel Hill: University of North Carolina Press, 1996.

Kulikoff, Allan. *The Agrarian Origins of American Capitalism*. Charlottesville: University Press of Virginia, 1992.

Lemon, James T. *The Best Poor Man's Country: A Geographical Study of Early Southeastern Pennsylvania*. Baltimore: Johns Hopkins Press, 1972.

Lord, Russell. *To Hold This Soil*. United States Department of Agriculture Miscellaneous Publication no. 321. Washington, D.C., 1938.

Lowenthal, David. *George Perkins Marsh, Versatile Vermonter*. New York: Columbia University Press, 1958.

Martis, Kenneth C. *The Historical Atlas of United States Congressional Districts*. New York: Free Press, 1982.

Mathew, William. *Edmund Ruffin and the Crisis of Slavery in the Old South: The Failure of Agricultural Reform*. Athens: University of Georgia Press, 1988.

McClelland, Peter. *Sowing Modernity: America's First Agricultural Revolution*. Ithaca, N.Y.: Cornell University Press, 1997.

McCoy, Drew R. *The Elusive Republic*. Chapel Hill: University of North Carolina Press, 1980.

McMurry Sally. "Who Read the Agricultural Journals? Evidence from Chenango County, New York, 1839–1865." *Agricultural History* 63 (Fall 1989): 1–18.

Meinig, Donald W. *Continental America, 1800–1867*. Vol. 2 of *The Shaping of America*. New Haven, Conn.: Yale University Press, 1993.

Merril, Michael. "Cash Is Good to Eat: Self-Sufficiency and Exchange in the Rural Economy of the United States." *Radical History Review* 3 (1977): 42–71.

Miller, Charles A. *Jefferson and Nature: An Interpretation*. Baltimore: Johns Hopkins University Press, 1988.

Miller, E. Willard, ed. *A Geography of Pennsylvania*. University Park: Pennsylvania State University Press, 1995.

Moore, John Hebron. *The Emergence of the Cotton Kingdom in the Old Southwest*. Baton Rouge: Louisiana State University Press, 1988.

Mumford, Lewis. *The Pentagon of Power*. New York: Harcourt Brace Jovanovich, 1964.

Nelson, Lynn A. "The Agroecologies of a Southern Community: The Tye River Valley of Virginia, 1730–1860." Ph.D. dissertation. College of William and Mary, 1998.

Novak, Barbara. *Nature and Culture: American Landscape and Painting, 1825–1875*. Rev. ed., New York: Oxford University Press, 1995.

Palmer, Laurence. *Fieldbook of Natural History*. New York: McGraw Hill, 1949. Rev. ed., 1975.

Parish, Peter J. "Daniel Webster, New England, and the West." *Journal of American History* 54 (Dec. 1967): 524–49.

Phillips, Sarah T. "Antebellum Agricultural Reform, Republican Ideology, and Sectional Tension." *Agricultural History* 74 (Fall 2000): 799–822.

Porter, Glenn, and Harold Livesay. *Merchants and Manufacturers: Studies in the Changing Structure of Nineteenth-Century Marketing*. Baltimore: Johns Hopkins Press, 1971.

Prince, Hugh C. "The Changing Rural Landscape, 1750–1850." In *The Agrarian History of England and Wales. Vol. VI, 1750–1850*, ed. G. E. Mingay. Cambridge: Cambridge University Press, 1989.

Prude, Jonathan. *The Coming of Industrial Order: Town and Factory Life in Rural Massachusetts, 1810–1860*. New York: Cambridge University Press, 1983.

Pyne, Stephen. *Fire in America: A Cultural History of Wildland and Rural Fire*. Princeton, N.J.: Princeton University Press, 1982. Repr., Seattle: University of Washington Press, 1997.

———. *Vestal Fire: An Environmental History, Told through Fire, of Europe and Europe's Encounter with the World*. Seattle: University of Washington Press, 1997.

Rezneck, Samuel. "The Depression of 1819–1822, a Social History." *American Historical Review* 39 (Oct. 1933): 28–47.

Roberts, Brian. *American Alchemy: The California Gold Rush and Middle-Class Culture*. Chapel Hill: University of North Carolina Press, 2000.

Rossiter, Clinton. *Conservatism in America*. New York: Knopf, 1955.

Rossiter, Margaret W. *The Emergence of Agricultural Science: Justus Liebig and the Americans, 1840–1880*. New Haven, Conn.: Yale University Press, 1975.

Rothbard, Murray N. *The Panic of 1819: Reactions and Policies*. New York: Columbia University Press, 1962.

Rubin, Jules. "The Limits of Agricultural Progress in the Nineteenth-Century South." *Agricultural History* 49 (April 1975): 362–73.

Ryder, M. L. *Sheep and Men*. London: Gerald Duckworth, 1983.

Savagian, John C. " 'Whether to the Haven of the Maelstrom?': The Rise and Development of Agrarian Reform Doctrine in Antebellum America." Ph.D. dissertation. Marquette University, 1998.

Scott, James C. *Seeing Like a State: How Certain Schemes to Improve the Human Condition Have Failed*. New Haven, Conn.: Yale University Press, 1998.

Sellers, Charles. *The Market Revolution: Jacksonian America, 1815–1846*. New York: Oxford University Press, 1991.

Shariff, Carol. *The Artificial River: The Erie Canal and the Paradox of Progress, 1817–1862*. New York: Hill and Wang, 1996.

Smith, Alfred Glaze. *Economic Readjustment of an Old Cotton State: South Carolina, 1820–1860*. Columbia: University of South Carolina Press, 1958.

Solbrig, Otto T., and Dorothy J. Solbrig. *So Shall You Reap: Farming and Crops in Human Affairs*. Washington, D.C.: Island Press, 1994.

Stewart, Mart A. *"What Nature Suffers to Groe": Life, Labor, and Landscape on the Georgia Coast, 1680–1920*. Athens: University of Georgia Press, 1996.

Stilgoe, John. *Common Landscape of America, 1580–1845*. New Haven, Conn.: Yale University Press, 1982.

———. *Borderland: Origins of the American Suburb, 1820–1939*. New Haven, Conn.: Yale University Press, 1988.

Stilwell, Lewis D. "Migration from Vermont." *Proceedings of the Vermont Historical Society* 5 (1937).

Stoll, Steven. *The Fruits of Natural Advantage: Making the Industrial Countryside in California*. Berkeley and Los Angeles: University of California Press, 1998.

Taylor, Alan. *William Cooper's Town: Power and Persuasion on the Frontier of the Early American Republic*. New York: Vintage, 1995.

Temin, Peter. *The Jacksonian Economy*. New York: W. W. Norton, 1969.

Thornton, Tamara Plakins. *Cultivating Gentlemen: The Meaning of Country Life among the Boston Elite, 1785–1860*. New Haven, Conn.: Yale University Press, 1989.

Trimble, Stanley Wayne. "Man-Induced Soil Erosion on the Southern Piedmont, 1700–1970." Master's thesis. Department of Geology, University of Wisconsin, Milwaukee, 1974.

Tyrrell, Ian. *True Gardens of the Gods: Californian-Australian Environmental Reform, 1860–1930*. Berkeley and Los Angeles: University of California Press, 1999.

Van Soest, Peter J. *Nutritional Ecology of the Ruminant*. Ithaca, N.Y.: Cornell University Press, 1982.

Weiss, Harvey. "The Origins of Tell Leilan and the Conquest of Space in Third Millennium Mesopotamia." In *The Origins of Cities in Dry-Farming Syria and*

Mesopotamia in the Third Millennium B.C., ed. Harvey Weiss. Guilford, Conn.: Four Quarters, 1986.

Welter, Rush. *The Mind of America, 1820–1860*. New York: Columbia University Press, 1975.

Williams, Raymond. *The Country and the City*. Oxford: Oxford University Press, 1973.

Wines, Richard. *Fertilizer in America: From Waste Recycling to Resource Exploitation*. Philadelphia: Temple University Press, 1985.

Wright, Chester Whitney. *Wool-Growing and the Tariff*. New York: Houghton Mifflin, 1910.

Zohary, Daniel, and Maria Hopf. *Domestication of Plants in the Old World*. 2nd ed. New York: Oxford University Press, 1994.

Index

Printed in the USA
CPSIA information can be obtained
at www.ICGtesting.com
LVHW091130150724
785511LV00001B/56

9 780809 064304